a guide for
adults with
hip dysplasia

by

Dr Sophie West
and
Denise Sutherland

A comprehensive book for adults with developmental
dysplasia of the hips, covering everything from hip
anatomy, diagnosis, and treatment, to hospital stays,
recovery, and getting back to normal life.

Includes over 400 quotations from hip dysplasia patients.

a guide for
adults with
hip dysplasia

by
Dr Sophie West
and
Denise Sutherland

sutherlandstudios

© Denise Sutherland and Sophie West 2011.
ISBN: 978-0-9872152-0-8

Designer, illustrator, and indexer: Denise Sutherland
Photographer: Sophie West

Photos of hospital locations used with permission of the University College London Hospitals

Set in Adobe Garamond Pro, Myriad Pro, and News Gothic

National Library of Australia Cataloguing-in-Publication entry (pbk)

Author: Sutherland, Denise.
Title: A guide for adults with hip dysplasia / Denise Sutherland, Sophie West
ISBN: 9780987215208 (pbk.)
Notes: Includes index.

Subjects: Hip joint--Dislocation, Congenital. Hip joint--Abnormalities. Pain--Treatment. Hip joint--Surgery. Postoperative care.

Other Authors/Contributors: West, Sophie.
Dewey Number: 616.72

Published by Sutherland Studios
ABN 67 766 872 800
PO Box 3849
Weston Creek ACT 2611
AUSTRALIA

Dedication

To each and every one of the wonderful Hip Women
and HipMen we are privileged to know.

You are an inspiration to us both.

This one's for you.

Other books by the same authors

Denise Sutherland

Word Searches For Dummies (May 2009)

Cracking Codes and Cryptograms For Dummies (September 2009)

Solving Cryptic Crosswords For Dummies (August 2012)

Cryptic Crosswords For Dummies (August 2012)

DISCLAIMER

Foreword

Being diagnosed with hip dysplasia in adult life often comes as a significant shock. This may be because for a long time symptoms may be reasonably manageable, and then the hip suddenly decompensates, and pain and disability become more of an issue. Often the condition is undiagnosed for a considerable period of time, with subtle abnormalities not being recognised on x-rays. When it comes to the realization that a major surgical procedure is required to correct the problem, this can be very daunting.

Unlike for total joint replacement surgery, there is much less information generally available regarding hip dysplasia and the surgical treatment of this. The relatively lengthy rehabilitation following surgery puts a lot of added stress on families and work commitments. This book provides much needed background information for those diagnosed with this condition, and provides considerable detail in terms of what can be expected during the hospital admission and rehabilitation afterwards. I am sure that being provided with such comprehensive information will make future patients feel less fearful of what lays ahead, and benefit their recovery and rehabilitation.

Mr Johan Witt

FRCS FRCS(Orth)

Consultant Orthopaedic Surgeon

The London Clinic

Abbreviations

◊ CAD: congenital acetabular dysplasia

◊ CDH: congenital dislocation/ dysplasia of the hip

◊ CBT: cognitive behaviour therapy

◊ CT: computed tomography

◊ DDH: developmental dysplasia of the hip

◊ DVT: deep vein thrombosis

◊ ECG / EKG: electrocardiogram

◊ ER: emergency room

◊ FAI: femoral acetabular impingement

◊ FHR: full hip replacement (same as THR)

◊ FO: femoral osteotomy

◊ GA: general anaesthetic

◊ GP: general practitioner (PCP)

◊ ICU / ITU: intensive care unit, intensive treatment / therapy unit

◊ IV: intravenous

◊ MRI: magnetic resonance imaging

◊ NBM: nil by mouth

◊ NHS: National Health Service

◊ NSAID: non-steroidal anti-inflammatory drug

◊ NWB: non-weight-bearing

◊ OA: osteoarthritis

◊ OS: orthopaedic surgeon

◊ OT: occupational therapist

◊ PACU: post-anaesthesia care unit

◊ PAO: periacetabular osteotomy

◊ PCA: patient-controlled analgesia

◊ PCP: primary care physician (GP)

◊ PE: pulmonary embolism

◊ PT: physical therapy

◊ PTSD: post-traumatic stress disorder

◊ ROM: range of motion

◊ THR: total hip replacement

◊ WHO: World Health Organisation

Notes:

◊ All doctors' and surgeons' names mentioned in patient quotations have been reduced to random initials, for anonymity.

◊ Surgeons are sometimes called Mr. or Miss, rather than Dr.

◊ All web sites mentioned in the text were accessed in November 2011.

Contents

Preface

After two years of hard work, here it is — the book we wished *we'd* had long ago! We are both 'hippies' with hip dysplasia, and after going through our own 'hip journeys', felt that it was important to develop a resource for others going through similar things. Until now, there has not been a single book published for adults with this condition.

This book is a comprehensive guide to developmental dysplasia of the hip in adults. We hope it will help you, whether you were initially diagnosed and treated as a child and have ongoing hip problems, or have developed the condition later in life, and are now having to deal with the shock of diagnosis. We've included everything we can think of, from basic hip anatomy and what hip dysplasia is, surgical options, coping with increasing disability, and the hassles of disability parking, to what to take to hospital, how to cope with bad room-mates, recovery from surgery, and getting back to normal life. We've written a chapter for carers, and one of projects you can make. There is a glossary to decipher the commonly used terms, a recommended reading list, references, and an index.

We met on the Yahoo Group HipWomen. This email support group (which has over 1,200 members, at the time of writing) proved to be an incredible resource for us, not only for the generous personal support we each received on our own hip journeys, but for discovering the problems and questions that real people with hip dysplasia have. This has given us a depth of insight beyond our own experiences. Many of the subjects addressed in this book are a direct result of conversations and topics that have been discussed by HipWomen members.

We have also been extremely lucky that 50 women and men have contributed over 400 of their experiences and thoughts for this book. We have appreciated each and every contribution, and we strongly feel that they will help you feel supported, whatever your situation. We want the book to be able to 'hold your hand' on *your* hip journey, and we wish you the very best. Have courage — it's a rough road, but it does finish, and generally with good outcomes.

Denise   Sophie

Acknowledgements

This book would be nothing without the input and hard-won experiences of these brave and wonderful hip patients (and a couple of husbandsand mums). Thank you each and every one of you for generously and candidly sharing your stories with us:

Adrian West, Alison Pepe, Alvaro de Lanzas, Annick Hollins, Arpine Babloyan, Betsy Miller, Brenda Kelley, Bri Keenan, Cathy Sanchez-Latin, Chrissie Grenfell, Christelle L. Smith, Claire Pierce, Clare Chisholm, Cynthia Keenan, Danielle Sturch, Dawn Hoggart, Debbie Sharpe, Deirdre Fulton, Emily Bell, Frances Phillips, Freja Swogger, Hannah Purdy, Ina Rosales, Jill Murphy, Jodi Seidler, John Swogger, Juan Carlos Vasquez Valencia, Katie Boyd Travis, Kristine Payant, Lara Lindeman, Laura Pokorny, Lea-Anne Hall, Lisa Jensen, Louise Jones, Marcie Pratt, Margaret Stott, Meghan Moya Woods, Melanie Anderson-Kong, Monica Shah, Rhianna Essex, Sally Jacober Brown, Sandra Taranto, Serena Zea, Shelley Gallagher, Shilpa Narayan, Sian Thomas, Teri Marsh, Tina Gambling, and Tina Louise Bayley

For their professional help on the book, our deep gratitude goes to:

◊ **Mr Johan Witt**, FRCS FRCS(Orth), Consultant Orthopaedic Surgeon, of The London Clinic

◊ **Dr Rob Reid**, MBBS, FACSP, FASMF, FFSEM (UK), Specialist Sports and Exercise Physician, of Canberra Sports Medicine

◊ **Dr Radha McKay**, BMedSc, MBBS (Hons), FANZCA, DipDHM, VMO at Royal Prince Alfred Hospital, Sydney, for her expert advice on anaesthesics

◊ **Monica Raven**, BSocSc.(Psych), GradDipApp.Psych., Assoc MAPS, for her expert advice on our 'emotional' chapters

Our thanks also to Jenny Sutherland (an ever-obliging model), Dr Charles Price and the International Hip Dysplasia Institue, Freja Swogger, Kal Starkis, Katrina Proust, and Sherrey Quinn. Most importantly, our thanks to Betsy Miller, Isobel Hannan, and Ralph Sutherland for their invaluable and painstaking work reviewing and editing the book, and Shirley Campbell for her expert editing of the index. Any errors that remain are ours alone!

Our personal thanks

DENISE

I would like to thank my husband Ralph, kids Rodger and Jenny, and mother-in-law Chris for looking after me before and after my hip replacement surgery. I would also like to thank Dr Rob Reid and Dr Catherine Rowe for being truly awesome doctors, Professor Wayne Southwick (New Haven) and Dr Damian Smith (Canberra) for being excellent surgeons, and David Berg for being such a lovely and skilled physiotherapist. And thanks to Mum and Dad for doing the right thing by my hip when I was little, even though it was traumatic for all of us.

SOPHIE

I would like to thank 'Westy', the most understanding and supportive husband in the world, without whom the last few years would have been unbearable. I would also like to thank my parents and sister for looking after me and supporting me through my hip journey, you have all been awesome. And of course, the many medical professionals who have helped at every stage, and shown care and courtesy to one of their colleagues.

Introduction

We thought you might like to know more about us, and how we came to write this book.

DENISE'S STORY

My hip journey has lasted for nearly my entire life. I was diagnosed with congential dysplasia of the (left) hip (CDH) in early 1966, aged around 18 months. It was picked up because I still wasn't walking. Right away I underwent a great deal of often very traumatic hospitalisations, treatments, and surgery in Australia, in both Canberra and Melbourne. I had a **tendonectomy** and two **open reductions** (see the Glossary for definitions of terms in **bold**) in less than a year, all of which failed. I developed **post-traumatic stress disorder** (PTSD) from my childhood surgery experiences (diagnosed as an adult). It was also pretty horrible for my poor parents.

In 1969 we moved to the United States. There we met **orthopaedic surgeon** Dr Wayne Southwick at Yale-New Haven Hospital. He tried some experimental techniques with my difficult hip. Firstly, I was put in traction with a metal pin through my knee. Then I had yet another open reduction, which rotated my left leg inwards, and finally, a few months later, another surgery to break and re-align my femur, which rotated the lower part of my leg outwards (while the hip joint was effectively rotated inwards into a more functional position). And it worked! After five major surgeries, eleven minor surgeries, 58 weeks in full body spica plaster casts, and six months in a sleeping brace, over a period of three and a half years, I had a functioning hip. I was 5 years old.

My hip was pretty well behaved for a long time. My left leg was a bit short, and rotated outwards, so I was always getting my foot trodden on, but otherwise things were fine. I was never much of one for sport, and just got on with my life. I was quite proud of the massive 'shark bite' scars down the side of my left leg.

We returned to Australia in 1975. I finished school, studied science, and then graphic design. My hip started to make itself known to me again in 1988. I had very painful pregnancies with my two children (born in 1988 and 1991), with worsening groin and hip pain. My first labour was obstructed and very long; my son was born by emergency Caesarian after

32 hours … x-rays later revealed that my pelvic outlet was quite deformed and no baby would *ever* get out that way. The pregnancies put a lot of stress on my hips. After my daughter was born by planned Caesarian in 1991, my doctor advised me not to have any more children, as my hip probably wouldn't hold up. I quite agreed!

In my early 30s, I started to get more groin and hip pain — **osteoarthritis** was kicking in. I had always known that my hip had a limited life, and would most likely need replacement when I was relatively young (before 50, most likely). I stuck it out for over ten years, until 2008 when my physiotherapist insisted that I saw a surgeon, I really didn't want to do it! PTSD really kicked in with having to revisit all this childhood hip surgery stuff. My poor old left hip was replaced in November 2009.

The surgery was quite tricky; my surgeon Dr Damian Smith had to order in the smallest hip **prosthesis** from overseas, as my femurs are very small. Afterward the surgery, I got a small **pulmonary embolism** and **obturator nerve** damage, which made recovery rather difficult and slow. Dr Smith was also able to lengthen my left leg, so for the first time I had even leg lengths! This took close on a year to adjust to, with a lot of **physiotherapy**, what with the nerves being stretched and so on.

I can attest that two years out from my surgery, my new hip is absolutely wonderful. While I do have permanent **hip restrictions** (my particular prosthesis has a higher risk of fracture), my hip is pain free for the first time in years, and my range of motion is vastly improved. My last surgery exactly fulfilled Dr Southwick's prediction of 40 years' wear from my reconstructed left hip — I was 5 when he did his last surgery on me, and 45 when it was replaced!

My hip replacement really marks the *start* of my adult hip surgeries (not the end of my dysplasia for good). The prosthesis will probably eventually need to be revised (replaced). And my right hip also has dyplasia, although not as severely. It will eventually need to be replaced too.

I have unfortunately also developed extensive osteoarthritis in both my feet and ankles. This has been put down to a lifetime of hip problems, walking on uneven-lengthed legs, limping, weak leg muscles, and so on.

I came across the HipWomen group on Yahoo about a year before my hip replacement; someone on another online group pointed me in their direction. It was truly wonderful to be able to communicate with other women who had similar stories to mine! As I'm an author, I thought it would be a great idea to write a book about hip dysplasia for adults (most information out there is about childhood or canine treatment), drawing on the incredibly vast accumulated knowledge of the HipWomen members. When I was starting to plan the book in earnest, Sophie joined HipWomen, with similar aims, and we quickly joined forces!

SOPHIE'S STORY

I qualified as a doctor in 2006 from Guy's, King's and St Thomas' School of Medicine, having wanted to be a doctor since I was 3, when I was given a plastic doctor's set to play with. My first two years of training quickly made me realise that I wanted to become a surgeon, and that I was developing an interest in **orthopaedic surgery**. I decided to complete an MSc. in Sports Medicine before starting my basic surgical training. It was during this period that I started to have problems with my hips. I'd had no hip problems during childhood.

I was diagnosed with hip dysplasia aged 27, at the start of my surgical training in 2009. Even with a medical background, I had no real idea what the diagnosis meant or the types of surgeries I would be facing — I thought it was something that babies had. I felt incredibly naive and a little bit embarrassed that my knowledge on the subject was so lacking.

When I started trying to find information about the condition on the internet — my first search produced results about dogs, not particularly reassuring — I realised very quickly that there was not a great deal of good quality information out there for adults. Due to being 'in the business' I was very lucky and got a diagnosis quickly, though I was in denial about the treatment. Only after the *sixth* specialist told me I needed a pelvic osteotomy did it start to sink in.

I initially had an **arthroscopy** to confirm the amount of damage to the joint on my left side followed by a **periacetabular osteotomy** on my left hip [see Glossary and Chapter 4 for explanations]. The recovery from this was long and painful, and complicated by a rare heart condition that I developed post-operatively. However, the biggest challenge for me was the emotional side; all I could think about was how I would have to go through this again with my right hip the following year. I have lots of friends who are doctors and a very supportive family, however none of them really understood what I was going through. This was my first real experience of truly understanding what being a patient is like, and it was eye-opening.

After this, I started to work at the same hospital where I was being treated. In my lunch times, I used to visit other hip dysplasia patients, to offer them support from my 'double role' as patient and doctor. It was during one of these encounters where I heard about the Yahoo Group HipWomen. I promptly decided to join. It was like a massive weight had been lifted off my shoulders — here was a group of people who knew *exactly* what I was going through. Just being able to talk to others who were in the same situation was such a relief. The second PAO was easier because of this support I was getting, bizarrely, from a group of women I had never met. I now consider many of these women as friends.

I love technology, I am usually one of the first to get the latest phone or iPad or gadget. However, during all this, a lot of the time I just wanted a good old fashioned book to be able to go to and look something up. I also wanted something to read in the hospital as reassurance.

Denise was already planning on writing this book, and when I joined HipWomen, she asked if I wanted to come on board. I felt it was such an important project, and would have been an invaluable resource for me, that I jumped at the offer.

I strongly believe the story of my hips, and my journey with hip dysplasia, has made me the person and surgeon I am today, and hope to be in the future. Looking back, I wouldn't change anything. It has given me opportunities to do and be a part of things that I would otherwise not have had or enjoyed. It is part of who I am.

Chapter 1
An Introduction to Hip Dysplasia

This chapter covers background information on hip dysplasia. This chapter does include rather a lot of medical jargon, which we have done our best to explain along the way. We feel it's important to include the correct medical terms, so you can achieve a greater understanding of the condition. It will also help you to have more effective two-way communication with your doctor. These words are included in the Glossary at the back of the book. The first instance of a word that is in the Glossary is set in **bold**.

What is hip dysplasia?

Let's start with the medical jargon — look out, there's plenty of it!

First of all, **dysplasia** means 'abnormality of development', so the term is applied to quite a few medical conditions, not just hips.

The condition of hip dysplasia has many names and abbreviations including: **congenital dislocation or dysplasia of the hip** (CDH), **congenital acetabular dysplasia** (CAD), **developmental dysplasia of the hip** (DDH), dysplastic hips, and so on. All of these names essentially refer to the same condition. DDH or developmental dysplasia of the hip is now considered to be the correct and standard terminology. It does not have the stigma that a **congenital** condition sometimes has attached to it, and covers the wide spectrum of severities that are seen, from relatively mild to severe deformity.

Developmental dysplasia of the hip (or DDH for short) is quite simply a failure of the hip joint to develop normally. This can range from a completely dislocated hip that is seen in babies, to the hip joint that just does not develop normally over time. It is this second group that is commonly present in adulthood. However, there are also many people out there who have had treatments for dysplasia as a child, but still have ongoing hip problems in their adult lives.

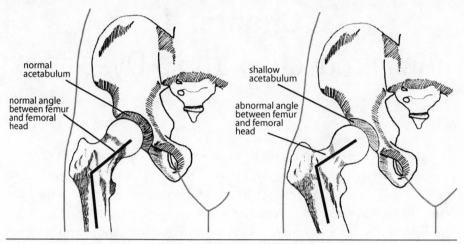

Figure 1.1 : Comparison between a normal hip (left) and a dysplastic hip (right)

As with all joints, stability comes from three major components: the bony structure, **ligaments**, and **muscles**. In the hip, the *acetabular labrum* (the **cartilage** rim around the socket) is also important. If you have a problem with any one of these components, it can lead to an unstable joint.

In DDH, the problem is with the bony structure. The *acetabulum* or hip socket does not develop into the nice curved shape that fits with the head of the **femur** (thigh bone). The socket is too shallow, with insufficient coverage of the *femoral head*, which means that the hip joint is trying to work with abnormal stresses on it[1]. Sometimes there can be an abnormal angle between the femur and the femoral head (although this is not present in all cases of dysplasia). See Figure 1.1. All of this leads to pain and, if allowed to progress untreated, to the development of **osteoarthritis**.

The labrum sometimes tries to compensate for the shallow bony socket, and can become enlarged (**hypertrophic**) to increase the relative size of the acetabulum[2]. Over time the labrum cannot withstand the excess stress placed on it, and can tear. Once **symptoms** have started to appear, it is usually a sign that the hip joint can no longer compensate for its abnormal structure. See Figure 2.3 on pg 17 for a diagram showing the acetabular labrum.

You might like to compare two x-rays, one of a normal hip and one with DDH (Figure 1.2). As you can see, reading x-rays is no easy task, and even doctors can miss the subtle differences between a normal hip and a dysplastic one.

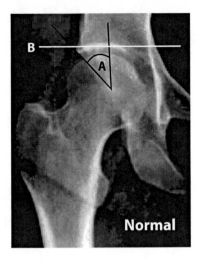
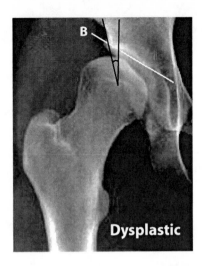

Figure 1.2 : Lateral central edge angles and weight-bearing zones. The normal hip (left) shows the normal angle (more than 25 degrees at A) The line B illustrates the weight bearing zone of the joint; the dysplastic hip on the right shows the lateral centre edge angle is reduced, and the weight-bearing zone is sloped.

There are two features to be aware of when looking at hip x-rays: the **lateral central edge angle**, and the **weight-bearing zone**.

The lateral centre edge angle is a measure of how deep the hip socket is. In a normal hip, this angle is *greater* than 25° (marked as A in Figure 1.2). In the dysplastic hip, however, the angle is *less* than 25°, and can even be negative numbers in severe cases[3]. A shallow hip socket means that a significant amount of the cartilage covering the femoral head is 'outside the hip joint', or uncovered.

The weight-bearing zone is marked as Line B in Figure 1.2 above. In a normal hip this weight-bearing zone of the hip joint should be flat. In a hip with DDH, this zone is sloped (not flat). This means that when you walk, the majority of the load is going through the edge or rim of the acetabulum, rather than through the stronger hip socket, where it should be. Over time this can lead to damage.

The egg cup analogy

If you put an egg in an egg cup, it fits nicely and doesn't wobble. However, if you then put the same egg on a spoon (which is shallower than an egg cup), the egg can move around within the spoon — it's unstable, and if shaken around too much will fall out! In this model, the egg cup and spoon are the hip sockets, and the eggs represent the femur heads. The egg cup represents a normal hip socket, and the spoon is a shallow dysplastic one. Luckily in adults, it is not just our bones that lead to a stable joint — there are ligaments and muscles holding things in place too, which is why in adult 'hippies' (people with DDH), the hip rarely dislocates.

How common is DDH?

The actual figures for the **incidence** of hip dysplasia in the population vary depending on which classification is looked at, and whether you look at screened or unscreened populations. A *screened population* means, in this case, that babies are checked for dysplasia routinely as part of their postnatal care. Figures will always be higher in a screened population, as *every* baby is being checked (not just those who present with hip problems). In 2009, a worldwide systematic review of *unscreened* populations estimated the **prevalence** of established hip dysplasia to be 1.3 people per 1,000[4].

Why wasn't it picked up when I was little?

This is a slightly tricky subject. There generally isn't a single right answer or reason, but here are some possibilities:

The human body is very good at compensating and adapting. If your dysplasia was not severe enough to cause your hip to dislocate as a child, you may only develop symptoms when your body can no longer compensate for your wonky hip. In many cases, this can be well into adulthood.

DDH can be difficult to diagnose, and no screening test is completely reliable. This, unfortunately, means that cases slip through the net. However, technological advancements in medicine have led to more medical professionals having access to the tools needed to diagnose the condition.

A more recent theory is that there are a proportion of babies who are born with undislocated or stable hips, but as they grow their hips become unstable, and can even dislocate. These children will not present as hip dysplasia patients until they are older.

Many people have been diagnosed as children and undergone treatments, however these fixes do not always work or last lifelong. In recent years, again due to advancement of medical technology and surgical knowledge, there are new options available to treat these people.

Why have I got it? Causes & risk factors

As frustrating as this may sound — especially after talking about medical advancements in the previous paragraph — it is not known why the majority of people with DDH have the condition. There are some features that *may* give you a predisposition to DDH (listed below), however, in many cases none of these are present.

The condition is more common in females, there is a 4:1 female to male ratio, the left hip is most commonly affected, and only 20% of cases are **bilateral** (occuring in both hips)[5].

RISK FACTORS

◊ Baby was breech presentation **in utero** (even higher risk if vaginal delivery as opposed to Caesarean birth)

◊ Birth weight of the baby greater than 5kg (11 lb)

◊ **Oligohydramnios** (too little amniotic fluid in the womb)

◊ Baby's position in the womb, and swaddling in infancy

◊ Prematurity

◊ Connective tissue **laxity**

◊ Decreased resistance of hip to dislocation[4]

HEREDITARY FACTORS

Studies of twins and families confirm that there is a **genetic** susceptibility to DDH, with a 12-fold increase in risk for first degree relatives[6]. This means that if you have a first degree relative (brother/sister/parent/child) affected by DDH, then you are 12 times more likely to have DDH yourself. This isn't as bad as it sounds, though. Putting this in perspective, in a condition that affects 1.3 people in 1,000, the risk for first degree relatives is increased to 15.6 in 1,000, which means you are still *far* less likely to have the condition than to have inherited it!

Diagnosis of DDH

IN CHILDREN

We're just going to give you a brief outline of how the diagnosis is made in children. An adult is usually able to tell their doctor where the pain or problem is, but babies cannot do this, so doctors rely on other methods. Most new babies undergo a health check at birth, and again at around 6 weeks — part of this is to check the stability of their hips.

In babies, the hip joint is predominantly cartilage, and therefore will not show up on an x-ray, but it can be seen with an ultrasound scan. Once the baby is older than 4½ months, the bone has developed well enough for any dysplasia to be seen on an x-ray.

Dysplasia is often picked up in toddlers, when there is a delay in walking, an unusual waddling gait, different leg lengths, or limping. We recommend the book *The Parents' Guide to Hip Dysplasia* by Betsy Miller for an in-depth discussion of hip dysplasia in children.

IN ADULTS

Diagnosis of DDH as an adult can be a very long, slow, and deeply frustrating process. It can take years to get a definitive diagnosis. It often isn't clear why you are having hip pain, and doctors may even doubt you. Not all scans will show clear results, and reading them correctly needs an expert eye — and hip dysplasia experts aren't all that common. We discuss these problems in greater detail in Chapters 2 and 10. Below are some of the symptoms and signs you might be having as a hippie.

Symptoms

A symptom is something which you, the patient, notices as a problem. Pain is by far the most common symptom of DDH, but it can vary in character from person to person. There are different patterns of pain depending on the underlying cause.

A sharp intermittent groin pain is commonly one of the first features. It is worse with activity, and relieved by rest. This can progress over time, leading to modification of your activities or exercise, and can eventually become a major problem, even when walking. Occasionally, there may be episodes of acute sharp pain that can almost make the hip 'give way', following which the hip can be sore for several days[7]. This pattern of symptoms is referred to as *acetabular rim syndrome*, and indicates damage is starting to occur at the edge of the acetabulum, which, in the dysplastic hip, is taking most of the load[7]. The joint can get stiff during periods of acute pain, however the **range of motion (ROM)** is usually preserved.

Occasionally, the first symptom of DDH is a sudden sharp, severe pain in the groin that immediately limits activity, but eases off over several days. This can then be followed by periods of 'catching' or a feeling of your hip giving way. The pain is usually reproducible with certain movements that the person can describe to their doctor. This pattern can occur if the labrum tears acutely.

If arthritis is already developing, there is also usually stiffness in your joint that does not get better or subside over time, in addition to significant activity-related pain.

Signs

A **sign** is something which a doctor will notice about your condition when they examine you, or something that shows up on a test or scan, which you may not be aware of. A common sign of DDH is pain when your hip is flexed to 90° and then **internally rotated** — this is due to the neck of the femur hitting, or **impinging,** on the damaged portion of the acetabulum[7]. If the labrum has torn, certain movements of the hip can cause pain, and sometimes clicking. These vary depending on the part of the labrum that is damaged.

If arthritis has already developed, the affected hip's range of motion will be less than it was previously. If the condition is **unilateral** (only in one hip), the range of motion will be less when compared to your other hip.

The impact of diagnosis

We discuss many of these things in more detail in later chapters, and we've devoted three chapters to the psychological aspects of hip dysplasia and diagnosis, as we feel it's so important. We're just touching on this topic lightly here. If DDH is a new diagnosis for you, a million and one things will probably be running through your head, and it may all seem completely overwhelming and terrifying. *This is a normal reaction.*

IF IT'S A NEW DIAGNOSIS

As you can see, getting a diagnosis of DDH is generally quite a shock to the system! These stories are typical of most patients' experiences :

The first time I was told I needed a periacetablar osteotomy (PAO), I was floored. I think my jaw hit the floor. I couldn't believe that I had dysplasia because I had "lasted" 38 years without ever having a hip issue. I was a little comforted knowing that it wasn't the fact that I was continuing to play soccer in a league at a relatively older age. That my pain wasn't caused by something I did made me feel better. I continue to be more concerned with the notion that my children could have dysplasia, and hope they don't end up having to have surgery like me. **Alison, 38 USA**

I was scared. I've never even heard of dysplasia before. But just from the sound of it, it seemed complicated. **Bri, 15 USA**

When Mr D told me I had 'severe hip dysplasia', I felt like I'd been told that I had cancer. I'd never had anything wrong with me all my life. I just sat and cried — as a marathon runner for the previous seven years, my whole world was coming crashing in on me. **Annick, 47 UK**

I saw the surgeon's mouth moving, but I couldn't hear any words come out. It was like a dream, in slow motion, I remember thinking to myself : "Why is he making this stuff up?'" Talk about denial. I *never* thought the pain I was having was from needing a new hip. I was 50 years old, and never knew I had dysplasia. **Jodi, 56 USA**

I felt confused that I had lived my life for 29 years without knowing I had hip dysplasia. I was terrified about the surgery, and frustrated that I had to stop dancing, and other activities. However, I was also glad to finally have an explanation and a solution for my groin pain. **Melanie, 29 USA**

I was surprised, but relieved. **Alvaro, 38 Spain**

I was absolutely gobsmacked. I kept thinking that I must have said something to mislead the consultant. I knew I had terrible pain in my hip, I had seen the x-ray and MRI results, yet still I felt it couldn't be that bad. I read all that I could, which wasn't much, as it all seemed to be about children or Alsatian dogs! It took me a couple of months to accept the need for surgery. **Dani, 42 UK**

When I was first told that I had hip dysplasia, I was almost relieved because I had been in pain for so many years. It felt good to finally have an answer. And I felt even better when I heard there was a solution, although learning about having a major surgery was a big shock. **Emily, 18 USA**

I was diagnosed with bilateral hip dysplasia and a labral tear at 20 years old. I had only ever heard of dysplasia in dogs and babies, and was curious to know why it hadn't been mentioned earlier. By that time, the cartilage deterioration had led to instability and severe pain. I was not prepared to hear that I needed surgery. I had always associated hip surgeries with older patients who needed a replacement. I was overwhelmed at first, but I began researching and soon learned that there were many people my age who were facing the same challenges. The more information I had, the more comfortable I became. Talking with women who had similar stories to me was very comforting, and prepared me for what was to come! **Jill, 20 USA**

IF IT'S AN OLD DIAGNOSIS

If hip dysplasia is an old diagnosis for you, you may still be feeling all of the above. You may have had ongoing hip problems since infancy or childhood, but also you may have thought you were 'fixed' once and for all back then.

I have always known I had hip dysplasia, and that there was a chance I would need a **total hip replacement** (THR) later in life. But when I was told I needed a THR at the age of 22, my first thought was "No, they must be wrong. That's what old people have, not 22 year olds!" All my friends said to me was "Oh yeah, my Nan had one of those." Hardly a comparison! I then felt I needed to find out everything I could about dysplasia, THR, and my surgeon. I wish there had been a book like this to help me then! **Rhianna, 28 UK**

I was shocked. When I was a child, I was sure my dysplasia was treated and gone for good. So when I started having pain, I didn't even connect the two. I thought I'd pulled a muscle, or something of that nature. So I went and had the x-rays. I was sitting in the doctor's office, waiting for the results, and looking at the hip **prosthesis** poster on the wall. There was a wall of distance between me and that poster. I looked at it from the healthy person's point of view, thinking: "Some people have serious hip problems. They end up getting a replacement. This happens."

And then the doctor came and told me I had dysplasia (what else is new), and that I probably needed surgery, but had to go to another doctor to get his opinion. She left to get his contact information. And I started looking at that poster again. It felt so different now, knowing that I might be one of these "people who have serious hip problems". I imagined that "thing" inside me and wanted to cry. I didn't know at the time what a **periacetabular osteotomy** (PAO) was, and tried to settle the "surgery" thought in my brain. **Arpine, 28 USA**

I always knew that I was destined for more hip surgery as an adult, that my repaired hip had a finite lifetime, and that a young-age THR was what was to come. Not a pleasant thing to always have looming ahead! But it was no great surprise when I started to get osteoarthritis in my 30s. **Denise, 47 Australia**

I'm 39 and was diagnosed with bilateral congenital dislocation of the hips (CDH) rather late, at age 3 in 1971. I went on to have surgery at ages 3, 6, and 10. I had no real problems again until my late 20s. My parents were told to expect that I would need hip replacements relatively early in life, but there was no mention of possible problems in my late 20s. I have since learnt that this is the classic time that hips start to deteriorate again. I was just pregnant with my first child when I saw the orthopedic surgeon, who said I would need a triple pelvic and femoral osteotomy, or a THR in five years' time. I was 31 at this time, but had been suffering increased hip pain for the previous two years. I waited until my baby was 1 year old, and went ahead with the osteotomy.

I was told it would take at least a year to get back to normal, which it certainly did. I found it incredibly hard to go through the surgery and recovery with a small child. The demands of picking up, walking, and other activities were difficult for me to do. I felt I missed out on doing many things with her when she was young. I've since had further surgery on my other hip, but am still having problems. The initial osteotomy was eight years ago now, but my hips have been a constant problem since then, so I'm now waiting to get a second opinion for the possibility of THRs.

If I had known that I might have had more hip problems in my 20s, at least I would have been able to decide the best time to start my family. I was totally unaware that I would be needing any more treatment until, say, my late 40s or 50s, and that it would be a hip replacement. It came as a real shock, as no one ever mentioned that my hips were getting worse. I had been going to the GP for years, since a teenager, and taking ibroprofen for aches and pains in my groin and so on, but still no mention that I might need further hip surgery — just endless trips to physio which never did any good.

My parents were just as shocked as me, as they were never told the long term outcome of having CDH and surgery. I think patients should be told what *may* be in store in the future so they can make informed life choices with this knowledge. **Clare, 39 UK**

Physical consequences

If you've been diagnosed with hip dysplasia, it's very likely that you've been having problems with your hips. The good news is it's very likely that something can be done about it. The bad news is, it's not going to be easy, whichever treatment options you and your surgeon decide are appropriate.

PAIN

Long term or short term, the pain from DDH can be debilitating. It may be **pre-op** (before an operation/surgery), **post-op** (after an operation/surgery), or ongoing. The important thing is to find something that gives you control over the pain, and be aware that what works for one person may not work for another. Trial-and-error is often the only way to find out. We go into detail on pain control in Chapter 3.

MOBILITY

Your mobility may vary greatly from day to day, gradually reduce, or just be poor from the start. Unfortunately, if DDH is causing you pain, mobility issues will become a factor sooner or later, even if they are intermittent. There is more on this in Chapter 5.

SURGERIES

Surgeries for DDH can range from minor to major procedures, with associated variations in recovery periods. Surgeries can usually be divided into those that are joint-preserving, and those where so much damage has already occurred to the joint, that some type of joint replacement is required. See Chapter 4 for in-depth coverage of the surgeries involved with DDH.

ARTHRITIS

It is believed that congenital or developmental malformations of the hip joint (including DDH) constitute an individual risk factor for premature degeneration[8]. In other words, if your hip isn't normal, you're more likely to get osteoarthritis. Often, the amount of arthritis or joint damage that has occurred by the time you are diagnosed will have an effect on what treatments will be best for you, and how successful they will be.

EXERCISE

It is possible to exercise with hip dysplasia, and it is important too. However, the type of exercise you choose may need to be modified and adjusted, depending on your symptoms. As a basic rule high-impact exercises (running, jumping, and sports that involve them) are bad for untreated dysplasia, whereas non-impact exercises (walking, yoga, cycling, swimming, Pilates) are usually fine, as long as you work within your pain limits.

There are two phrases to bear in mind: "Some is better than none (however little)", and "Do what feels right for you." It is important to discuss your exercise goals with your surgeon before surgery, to find out what is realistic — each person is different. See Chapter 5 for more details.

Social consequences

WORK AND STUDY

The key feature with work is to be realistic and honest with both yourself and your employer. If surgery is on the cards, you will definitely need time off work. The amount depends greatly on your type of work and the procedure you have, but you will need at least six weeks, and possibly as long as three or four months. We include here stay-at-home parents, because that's a full time job too. When you are considering time off after surgery, it's important to factor in that you will need significant help at home. This is covered in more detail in Chapters 4, 9 and 16.

If you are a full time student, you will need to discuss your situation openly and honestly with your educational institution's administration and teachers. If you need surgery, you will need time off, probably in the order of six weeks to three or four months, depending on what surgery you have done. Your educational institution is there to help you work out the best outcome, and they may have options available that you haven't even considered. You may be able to study part time, or submit work online. You may like to defer for a semester, and pick up your studies again once you've recovered from your surgery. You might also be able to schedule surgery during the summer holidays, to minimise the impact on your classes. Keep in mind that being on crutches makes it almost impossible to carry textbooks, and slow to get around campus.

I was diagnosed in the spring, and had my surgery in that summer, only a couple of months later. This decision was made because it fit best with my schedule for college [university]. I didn't want to wait a few years, because I didn't know what I would be doing in the future. We also choose to get it done so soon because I was still under my parents' health insurance, so they could help pay for the surgery. Also, my doctor said that my condition would only get worse, so we wanted it done as soon as possible. **Emily, 18 USA**

FAMILY AND FRIENDS

A social support system is important for everyone, even if there are no medical problems to consider. Having family and friends around will make some of the low times easier. Their support will also be vital after any surgery you decide to have. This subject is covered in Chapter 9.

Keeping up with going out or socialising can sometimes be difficult, especially when you're on crutches or a walking stick. "Clubbing on crutches" isn't impossible, though! More on this in Chapter 6.

If you already feel completely overwhelmed, please don't worry, it's completely normal. Make yourself a cup of tea or coffee (or maybe a stiff drink!) and breathe. We've been there and done that ourselves, and are here to tell the tale — along with the dozens of others who have contributed their stories to this book, and thousands of other patients around the world. You will be *fine* — yes, it may take some time, and be a hard road ahead, but there is every reason to expect that you will have a good outcome at the end of it all.

Chapter 2
Hip Anatomy and Imaging

This chapter is divided into two sections. The first section provides information on the anatomy of the body that is important to hip dysplasia; the pelvis, hip joint, and surrounding muscles. The second section covers the medical imaging that can be used to find out what is wrong with your hip, and allow your surgeon to find out more information about the anatomy of your hip.

Hip anatomy

We admit that this section is very technical, with a lot of long names for muscles, **tendons**, nerves, and bones, and may be confusing. It isn't essential to read, so you can skip past it to the section on imaging if you like. But don't be put off — if you want to be able to better understand what your surgeon and physiotherapist are talking about, then this information will help you. Figures 2.1 – 2.5 will help you follow our discussion.

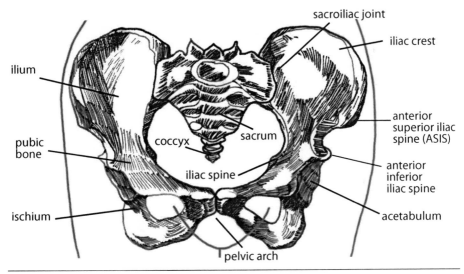

Figure 2.1 : The bony structure of a normal female pelvis.

HIP BONES AND JOINTS

The **pelvis** is an irregular ring structure. It is connected to both the spine (at the *sacrum*) and the thigh bones (the *femurs*), and allows transmission of the weight of the body to the legs.

The pelvis is formed from three bones, that are separate at birth and through childhood. As we enter adulthood, these bones fuse together to form a rigid ring structure. These are called the *ilium, ischium,* and *pubis,* and they meet at the *acetabulum* or hip socket. The *femur* is the long bone of the upper leg, and has two joints, at the knee and the hip. The head of the femur is at the hip end of things, which articulates (forms a joint) with the acetabulum. The femur's neck attaches to its shaft.

At the junction between the neck and the shaft of the femur are the *lesser* and *greater trochanter* (Figure 2.2). These are bony bumps where important muscles are attached. The greater trochanter is the bony bump you can feel on the outside of your thigh.

The hip joint is a ball-and-socket type of joint. It is formed by the acetabulum (socket) and the head (ball) of the femur (thigh bone), both of which are lined with a layer of *articular cartilage*. A **joint capsule**, which is a tough band of ligaments surrounding a joint, attaches to both the femur and the bony acetabulum, creating the **synovial** joint space. This is filled with fluid, which acts as a lubricant and a physical buffer allowing smooth movements of the joint. This cartilage and fluid act as shock absorbers for the joint.

Another feature of hip anatomy that is important we understand in DDH is the *acetabular labrum* (see Figure 2.3). It's name is often shortened to just *labrum*. This is a rim of cartilage (like flexible plastic) that surrounds the acetabulum. It makes the socket deeper, and holds the femoral head in the joint, making it more stable.

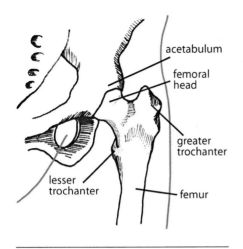

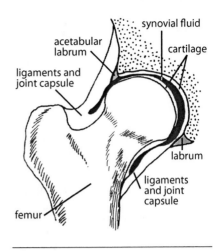

Figure 2.2 : A normal hip, showing how the femoral head fits snugly into a normal acetabulum.

Figure 2.3 : Normal hip showing the acetabular labrum.

MUSCLE GROUPS

The Hip Flexors

The hip **flexor** muscles flex and bend your hip. They include the ***iliacus, psoas major***, and ***rectus femoris*** muscles. The *iliacus* and *psoas major* muscles merge together (often jointly referred to as ***iliopsoas***). At the lower end they insert into the *lesser trochanter* of the femur. The *iliacus* muscle comes from the *iliac fossa* (the large curved surface of the internal surface of the ilium). See Figures 2.1 and 2.4. The *psoas major* muscle comes from the edges of the lumbar vertebrae (lower back/spine). *Rectus femoris* is one of the **quadriceps** muscles of the thigh. At the pelvis, it attaches to the front of the *iliac spine,* and just above the hip socket. The lower end of the *rectus femoris* attaches through the knee cap, and then into the **tibia** (shin bone).

The Adductors

The group of **adductor** muscles **adduct** your leg (pull it in *towards* the midline of your body). They include *adductor brevis* (shortest), *adductor magnus* (largest), and *adductor longus* (longest). They attach from the *pubis* (the bony bit you can feel in the front and centre of your pelvis) and the inside of the shaft of the femur. See Figure 2.4 on the next page.

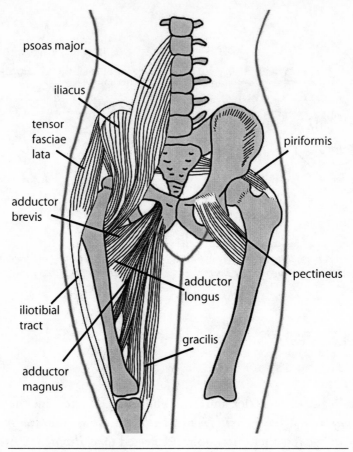

Figure 2.4 : Front view showing important muscle groups of the pelvis.

The Abductors

The *tensor fascia lata,* **gluteus** *medius,* and *gluteus minimus* muscles are **abductors** — they **abduct** your leg (pull it *away* from the midline of your body). They also help with the rotation and stability of your pelvis on the femur. The *tensor fascia lata* attaches from the *anterior* (front) *superior* (higher) *iliac spine,* and travels down the outside of the leg. It inserts into the *iliotibial tract* (a thick **tendon**), which then attaches to the outer edge of the top of the *tibia* (shin bone).

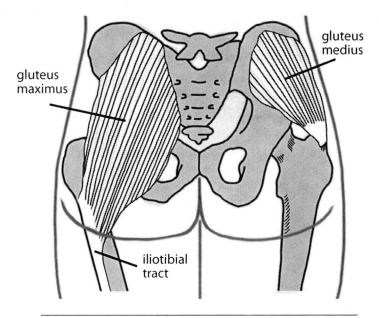

Figure 2.5 : Back view of pelvis showing the gluteus maximus and medius (gluteus minimus is hidden underneath gluteus medius).

The Gluteal (bottom) Muscles

These include *gluteus maximus, gluteus medius,* and *gluteus minimus* (which, as you can guess, are large, medium, and small in size). See Figure 2.5. The actions of these muscles change depending on whether your hip is straight or flexed (bent). In the straight position, the *gluteus medius* and *minimus* abduct your leg (move it away from midline), and stabilise the pelvis on the femur when walking.

The front part of *gluteus medius* and the *gluteus minimus* internally rotate the hip. However, when the hip is flexed, they act to *externally* rotate the thigh (along with *gluteus maximus*). Confusing, we know! *Gluteus maximus* is the largest muscle in the body (and what forms the curviness of your bottom cheeks). *Gluteus maximus,* and most of *gluteus medius,* also stabilise your pelvis on your femurs, especially when you stand on one leg. *Gluteus maximus* extends the hip.

The hamstring muscles are at the back of your leg, they also act to extend the hip.

NERVES AND BLOOD SUPPLY

There are many nerves and vessels that supply the hip joint, and all the surrounding bits and pieces. The main nerves we're concerned with in hip dysplasia are the **sciatic nerve**, *femoral nerve, obturator nerve*, and *lateral cutaneous nerve*. See Figure 2.6. The main *femoral artery* travels from deep within the hip, down the leg to the knee. We'll talk about the important nerves later, in Chapter 4, in relation to surgery and complications.

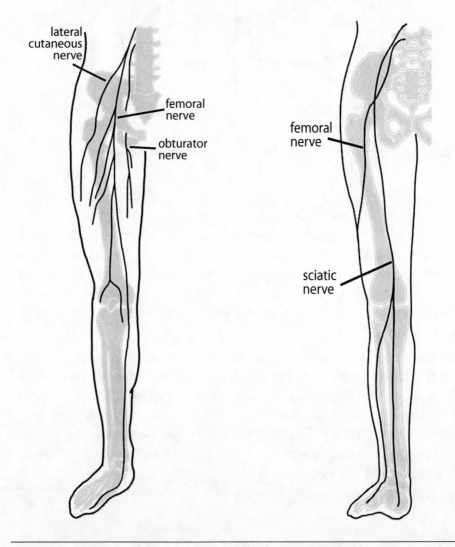

Figure 2.6 : Major nerves of the leg. Anterior (front) view on the left, posterior (rear) view on the right

Imaging methods

You will come across a range of imaging procedures frequently during your hip journey, as they are the only way to find out what's going on inside your hip without actually cutting you open. Thank goodness for modern technology! Note that for all scans, you will have to discuss the results with your doctor or surgeon; the people who *perform* the scans on you are usually unable to give you results.

EXPOSURE TO RADIATION

The amount of radiation that you receive from an **x-ray** or **CT scan** is minimal. It is calculated relative to the amount of background radiation that we are naturally exposed to on a daily basis, as part of normal life from natural sources on the earth. A pelvic x-ray is the equivalent to 1–10 days of background radiation, and a pelvic CT scan is equivalent to 2–3 years[9]. However, the actual risk of developing a condition as a result of exposure to radiation from medical diagnostics is *extremely* small. **Magnetic resonance imaging (MRI)**, as the name suggests, relies on magnetism to obtain the images, and therefore has no radiation exposure associated with it.

PLAIN X-RAY

An x-ray is one of the quickest and simplest diagnostic tools available. It allows the internal skeleton to be seen in two dimensions (2D). It does not allow accurate visualisation of any of the soft tissues. It can be used to see problems with the bony structure of the hip, pelvis, and femur. It is quick, and can often be done the same day as it is requested. It usually involves lying on a special surface (sometimes it is done standing against a special screen), while a technician takes the picture with an x-ray machine. The procedure is painless, although sometimes it can be a bit uncomfortable if the

HIP TIP : Sometimes you'll discover that your x-ray has shown up more than you intended - a genital piercing, a tampon, an IUD ... horror! While it may feel horrendously embarrassing to you, rest assured, all your medical staff, from the radiologist to your surgeon, have seen it all before, and won't even notice. Please don't worry!

technician needs to you hold your leg still in an awkward position. They have foam bolsters which can help hold your leg in a particular position. X-rays will show dysplasia, and can help quantify the degree of dysplasia, femoral structure abnormalities, arthritis (narrowing of the joint space), cysts in the bone, and **osteophytes** (extra bone at the joint edges)[10].

COMPUTED TOMOGRAPHY (CT) SCAN

A **computed tomography**, or CT, scan can give a three dimensional (3D) view of all the internal structures of your body. From an orthopaedic perspective, it is not so good at seeing all the soft tissues around a joint (an MRI is better for that). However, it is very good for looking at bony structures, and can also show the internal structure of the bones. As it producs a 3D image, far more detail can be seen compared to an x-ray. To have a CT scan, you lie still on a special bed that slides into a short cylindrical tunnel (imagine a ring doughnut) — see Figure 2.7. It takes only a few minutes for the scan to be done, and it's painless. A CT scan can be particularly useful to your surgeon in planning for your operation[10].

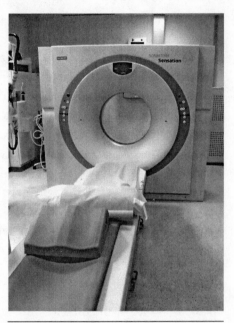

Figure 2.7 : A CT scanner. The table moves the patient in and out of the scanner.

MAGNETIC RESONANCE IMAGING (MRI)

An MRI scanner looks almost identical to a CT scanner (see Figure 2.7 above), but with a longer cylindrical tunnel (roughly 1.5 metres / 5' long). It uses alternating magnetic fields to visualise all the internal structures of the body. This is very good for looking at the ligaments, cartilage, muscles, and tendons around your hip joint. It can highlight areas of inflammation. An MRI scan will also show any signs of arthritis and occasionally **loose bodies** (fragments of bone or cartilage floating within the joint), and cartilage tears (though these are best seen with an **arthrogram**).

It takes between thirty minutes to an hour for an MRI scan. You lie on a narrow table, which is then moved inside the narrow tunnel. Your legs will probably be strapped together to keep them still. The tunnel is long and quite narrow, about 70 cm / 28" across, with the roof close to your head. Some people find this **claustrophobic**. The machine also makes very loud clunking, thumping, clicking, and buzzing noises. You will be given ear plugs at the very least, or possibly a set of head phones, which not only block out some of the noise, but also allow the technical team to talk to you throughout the scan. You can often listen to the radio through these head phones, and you might be able to get your own MP3 player hooked up in the control room — check with the technicians beforehand.

Most MRI scanners blow a light breeze down the tunnel, to help ease the sensation of being enclosed, and there are lights inside the tunnel.

MR ARTHROGRAM

An arthrogram involves having a special dye or contrast injected directly into the joint, and then the joint being scanned by an MRI machine. The injection into the joint is usually done under the x-ray guidance (**fluoroscopy**). Another dye that shows up on x-rays is injected first. Then your hip is viewed as a real-time moving x-ray, to make sure that the injection of the MRI contrast dye into your joint goes into exactly the right place. The injection can sometimes be uncomfortable, and you may feel pressure in the joint, but it's usually not too bad.

Local anaesthetic is sometimes used. This has two roles. Firstly, it can numb the joint, so the procedure is less uncomfortable. Secondly, it can also be diagnostic. If the inside of the joint is completely numb, and the symptoms persist despite this, it may indicate that the problem with your hip is **extra-articular** (outside the joint capsule). On the other hand, if all your symptoms abate when you have the local anaesthetic, then it is extremely likely the symptoms are caused by something **intra-articular** (inside the joint capsule). Following the injection of the contrast dye, you then have an MRI scan. This allows greater visualisation of the joint, particularly loose bodies, tears in the cartilage, and the acetabular labrum.

What to do if you are claustrophobic

Here are a few tips and tricks for coping with an MRI scan:

◊ Tell the doctors organising the scan if you anticipate any problems.

◊ Keep your eyes closed during the scan.

◊ Listen to relaxing music, an audio book, or meditation tracks during the scan. The strong magnetic fields generated by the MRI scanner would wreck your MP3 player if you had it on your body, but the technicians may be able to connect your player to their stereo system for you, so you can listen to it via headphones. Call ahead of time to see whether or not they can do this for you.

◊ Use relaxation techniques before and during the scan.

◊ Arrange a 'bail out' signal with the radiographer doing the scan; sometimes having the control to stop in your hands is enough to get you through. You will be given a 'panic' button in case you need to contact the technician during the scan.

◊ Arrange to have a mild sedative before the scan (you will need to be accompanied home if you have this).

◊ **Cognitive Behavioural Therapy (CBT)** — this can help you to overcome fears, but usually takes months.

If claustrophobia is a serious problem for you, and none of the above strategies are likely to help, find out if there is an 'open' scanner you can go to. An open scanner is specially designed for those with claustrophobia, and is not an enclosed tunnel shape. These machines are becoming more popular, as patients find them much less stressful. An open scanner is usually more expensive, and doesn't produce as high-quality images, though. Ask at the hospital imaging department, or your local private imaging companies, if they know of any such open MRI scanners in your local area.

Here are some experiences and ideas from several 'hippies' :

I am claustrophobic, so I asked for a Valium prior to the procedure.
Serena, 39 USA

I'm ever so mildly claustrophobic — I managed because the tech doing my MRI used a 'rear-view mirror', so I could see him the whole time. They also played music that I selected, so I was a bit distracted. I kinda have to 'talk myself off the ledge', and then I'm okay. **Alison, 38 USA**

I did feel very closed-in in the MRI machine, feet tied together, nose inches from the surface. I knew it would be like that, so in advance I burned a CD with my favourite pop songs — I figured that three-minute pop songs would divide the time up into little chunks so I could tell how much time had passed. My sister is really claustrophobic, so when she had an MRI, the staff placed her feet-first in the machine with the entrance in line with her nose, so she could see outside the machine, so she felt fine. **Freja, 45 UK**

I don't really like being stuck in small spaces, but the hip MRI did not seem that bad, because part of my head was sticking out of the MRI machine.
Melanie, 29 USA

I am claustrophobic, and was extremely panicky when going into the MRI machine — all I wanted to do was get out. I was given headphones with music, but with the noise of the machine, it was hard to hear it. I think that concentrating on my breathing, keeping my eyes closed, and thinking about other things, such as what I was making for tea kept me calm. Towards the end I actually started reciting the times tables! Maybe strange, but I'm not very good at them, and practising helped to keep me calm. **Sian, 29 UK**

We hope that you have found the information in this chapter useful. We appreciate some of it is very technical (especially the anatomy), but we feel it is important to include it so you have as much information at your fingertips as possible, and to use for reference.

There is a PDF of the basics of hip anatomy up on our web site <sutherland-studios.com.au>, which you can download, and use for reference, especially when you're at medical appointments.

Chapter 3
Non-Surgical Options

So, you have been given a diagnosis of hip dysplasia (DDH) — what next? Well, the good news is that there *are* treatments out there to help. It is important to realise that each case of DDH is different, and what may be suitable for one person is not for another. Some cases of DDH are milder than others, too. This chapter gives an overview of what non-surgical treatment options are available, and general information about each one.

Watchful waiting

Just because you have been given the diagnosis of hip dysplasia, it doesn't mean surgery is an absolute essential. No-one can make you have a surgical procedure unless you agree, and your dysplasia may be relatively mild. Having surgery is a major decision to make, with a long rehabilitation process. You must be completely committed and happy with your decision. DDH isn't a life or death situation, rather it's about your quality of life. If you feel you can manage with your situation at the present time, and are able to live your life as you wish, then you may decide not to put yourself through major surgery.

There are several factors to consider:

◊ Just because you do not feel surgery is right for you at this time, it is not a final decision. You may change your mind as the situation with your hips, and what you want from life, changes.

◊ People with hip dysplasia are more prone to develop osteoarthritis due to the abnormal joint shape. Several of the surgical options are about joint preservation (periacetabular osteotomy in particular), and the results can be better the *less* damage there is to the joint. Once arthritis has already developed, your options may become limited. Unfortunately, there is no way to predict exactly when arthritis might start in any given hip.

◊ You may already have significant arthritis. In this situation, the results of joint preservation surgery (such as PAO) are likely to be short-lived. You may then prefer to hold off as long as you can, until needing a total hip replacement (THR).

Change of lifestyle

There may be things you can change that can minimise the stress on your joints or make the pain easier to live with. The "What you can do to help your condition" section of Chapter 5 will also help.

EXERCISE

Exercise is really important for joint and whole body health. However, some forms of exercise put far greater stress on your hips than others. Running on hard ground is an example of high-impact exercise. Changing to running on a treadmill is one step down, but is still rather high-impact.

> HIP TIP : If pain-relief medication works well for you, make sure you have some on board before you start an exercise or physiotherapy session, but do not use it to 'push through the pain.'

You could consider taking up non-impact physical activities, for example cycling, using an elliptical trainer, cross country skiing, or swimming. Finding ways of exercising that reduce joint loads is important. Avoid repetitive-movement exercises (where a few muscles make the same movement over and over again for a long time). Start your program slowly, working within your limits of range of motion and pain, and gradually increase the level or intensity. You may like to consult with a physiotherapist / physical therapist, exercise physiologist, or sports physician for their expert advice.

What's helpful to realise is that no matter how bad your joints are, there will always be *something* that you can do. While finding the motivation can be difficult, especially when you're in pain, the more you can do (within your pain limits), the better. The human body was designed to move and it can get quite unhappy if it doesn't. An interesting example is if you put a completely normal knee in a plaster cast for six weeks, when it comes out it will hurt to move (even though there was nothing wrong with it). The more you can keep your joints moving the better. Building 'incidental exercise' into your day can help. This includes things like walking around when talking on the phone, cleaning the windows, walking the dog, or getting off the bus a few stops early, and parking further from your office.

Weight loss is an effective way of physically reducing the load going through your joints every day, this is discussed in detail in Chapter 5, and obviously any exercise you can do will help with this, if it's your goal.

SMOKING

Smoking increases the risk of infection post-operatively, and of respiratory complications post-anaesthetic. It also impairs healing. Even if you are not considering surgery at this stage, stopping smoking now will put your health and your hips in a far better position if you do decide to have surgery at a later date.

A PERSONAL DECISION

Betsy has written a very comprehensive analysis of why she has decided against hip surgery as an adult, so far:

I haven't had surgery as an adult. As a baby, (under 1 year old), I had traction, **closed reduction**, a spica cast, and a brace. When I first presented with hip pain as an adult, I decided to wait for surgery. The first orthopedic surgeon I went to was useless. Basically, he was pessimistic about the possibility of making any positive changes. He said I would need a THR in five to 15 years, and there was no way to mitigate the situation. He looked at my x-rays and said "You don't have much support on your left side. How are you walking around?"

After pushing hard, I was able to see another orthopedic surgeon, Dr G, who specialized in hips. After seeing him, I delayed surgery based on these considerations:

1. My cartilage looked good, and I did not have arthritis.

2. Because of congruency issues with the shape of my socket and non-round femoral head, I was not a good candidate for a PAO.

3. I had young children and a husband who had to travel a lot for work. Both of my parents were deceased, and my mother-in-law was in poor health. It would have been very hard to get support at home.

4. My job as a writer was sedentary, so I wasn't faced with the problems I would have if I was a waitress or nurse, for example, who has to be on her feet all day.

5. Since I was under 40, and looking at a THR or nothing, it was best to wait and try to minimize the number of **revisions** (replacement THRs) I would need in my lifetime. At the time I was seen, the best estimate was 15 years before a revision would be needed.

6. My hips had been hurting less than a year when I was seen, but it did make sense to wait and see if my hips would stabilize, which they did after about a year of chronic pain. Now I just have to pay attention to how far I walk, and I can typically manage pain with over-the-counter ibuprofen.

So I would say find a doctor who is willing to explain the risks/benefits of surgery in your specific case. Pay attention to how your hips are doing over time and to your quality of life. **Betsy, 48 USA**

Physiotherapy and hydrotherapy

Figure 3.1: Hospital hydrotherapy pool.

When we have pain in a joint, the muscles around that joint sometimes 'switch off'. This can lead to weakness, less stability around a joint, increase in pain at that joint, and sometimes pain in other joints due to muscle imbalances. Knees are particularly prone to this if you have hip problems. **Physiotherapy/ physical therapy** can help you learn to get these important muscles working better, including your "core stability" muscles. It can teach you ways to exercise that reduce the stress on your joints. This can often lead to pain reduction.

Hydrotherapy is a type of physiotherapy that happens in a special very warm swimming pool (Figure 3.1). Exercising or physiotherapy in water is great for joints, as the water supports the weight of your body. The warm water helps relax tight or weak muscles, allowing them to work better. The water also provides gentle resistance to your muscles.

A Pilates program and yoga are both forms of exercise that concentrate on balance, flexibility, and core strength. As they are low or even non-impact activities, they are unlikely to be doing any damage to your joints. They can also be good for those who are **hypermobile** (having an unusually wide range of freedom of movement or flexibility in a joint).

With all of these treatments, it is important to realise that although they can help with symptom control, and help you to lose weight (thereby easing the load on your joints, and hopefully slowing the progression of damage), they are not addressing, and *cannot* address, the underlying structural abnormality that exists in DDH. Having said that, symptom relief is a very important factor, we know first-hand how miserable life can be when you are in pain, or your joint movement is restricted.

Analgesia (pain relief)

Analgesia is the medical term for pain relief. We are going to cover some of the commonly used analgesics, or pain killers, that can be taken orally. This is *not* a definitive list and you may find you have been prescribed something or advised something that we have not covered. The names for most medications vary between countries, as do the licensing laws for

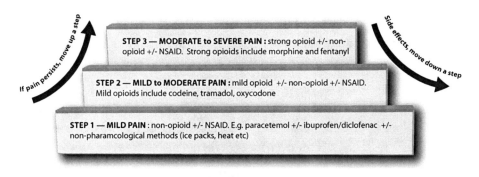

*Figure 3.2: Analgesic ladder — adapted from the **WHO** analgesic ladder.*

drugs. This can mean that something that is available over the counter in the UK may only be available on prescription elsewhere. If you have any doubts or concerns, discuss them with your doctor.

Before we go into the details of the various types of pain medications, we want to share the idea of an 'analgesic ladder' with you (Figure 3.2). The idea is that you start at the bottom (Step 1). If, after trying that medication, the pain persists, you move up a step. Like with a real ladder, it's important to start on the bottom rung. If you get too many side effects or problems with drug toxicity, you move back down a step.

You may find analgesic creams, like Voltaren, helpful, and they are certainly worth a try, but in general the source of pain (your hip) is too deep to be affected by such topical creams.

SIDE EFFECTS

Every drug has its own side effects and, while some are more common, others are incredibly rare. We're only going to mention some of the common side effects; before you take any medication, it's important to read the information dispensed with it. There can also be increased risks of side effects with long-term use of some medications. If you think you may be experiencing any adverse reactions, seek advice from a medical professional. Your pharmacist is also a very knowledgeable source of drug information.

PARACETEMOL

◊ **Examples**: Acetaminophen, Atasol, Calpol, Efferalgan, Febridol, Feverall, Panadol, Panadol Osteo (slow-release), Panamax, Triaminic, Tylenol.

◊ **Effects**: Paracetamol has analgesic and **antipyretic** properties (it helps reduce pain and fever). It is a mild pain killer. It can be combined with other pain killers to gain a greater pain relieving effect.

◊ **Side Effects**: It is usually well tolerated with few side effects when taken within the recommended guidelines. The main adverse effect is liver damage if an overdose is taken.

◊ **Contraindications**: Hypersensitivity/allergy to paracetamol, liver damage (in high doses).

NON–STEROIDAL ANTI–INFLAMMATORY DRUGS

This wide group of medications (abbreviated as NSAIDs) includes many varieties that all work a bit differently. The main thing that unites them is that they are *not* steroids. They are used to reduce pain, fever, and inflammation.

◊ **Examples**: Advil, Anadin Joint Pain, Arthrotec, aspirin, Brufen, Caldolor, diclofenac, Dicloflex, ibuprofen, Motrin, naproxen, Nurofen, Voltaren, Voltarol

◊ **Effects**: NSAIDs do exactly what their name suggests — reduce inflammation — and by this action they reduce pain. They also have an anti-platelet effect, meaning that they can increase bleeding times. They can be taken with paracetemol.

◊ **Side Effects**: Nausea, indigestion, gastrointestinal ulceration, bleeding, raised liver enzymes, diarrhoea, constipation, nose bleeds, headache, dizziness, rash, salt and fluid retention, and high blood pressure.

◊ **Others**: There are other types of NSAIDs available that work in slightly different ways, eg Celebrex. It is always best to discuss your individual needs with the doctor who prescribes them for you.

If you find that you get a lot of relief from taking anti-inflammatory medication, but are worried about the long term consequences of gastrointestinal problems, it is worth discussing this with your GP. There are several pharmacological options available that can protect your stomach. Taking NSAIDs at the end of a full meal, or with an antacid, can also help to reduce irritation of the stomach.

OPIOID–BASED DRUGS

Opiods are strong pain killers that are derived from morphine-based substances. You are more likely to be prescribed these after surgery, but if your condition is very severe, you may need these before surgery.

The strength and preparations can be variable; codeine preparations are some of the weaker ones, with tramadol and oxycodone being stronger. They are only available with a prescription, and are controlled substances in some countries.

◊ **Examples:** Codeine, codeine phosphate, dihydrocodeine, Dihydrin, Dromadol, Endone, Mabron, Norco (hydrocodone with paracetamol), oxycodone, Oxynorm, OxyContin, Panadeine (codeine phosphate with paracetemol), Percocet (oxycodone with paracetemol), Percodan, Ralivia, Roxicodone, Ryzolt, tramadol, Tramahexal, Tramake, Tramal, Tramedo, Ultracet (tramadol with paracetamol), Ultram.

◊ **Side Effects:** The risk of side effects increases as the dose increases, emphasising the importance of only taking enough to get adequate pain relief. Common problems include drowsiness, nausea, dizziness, and constipation. Occasionally, these can be severe enough to stop people wanting to take the drugs. If you need to take opioid analgesia regularly, you may need to consider taking a **laxative** simultaneously to avoid the effects of constipation. Also read our tips for dealing with constipation on pg 247.

It is possible to become tolerant to these drugs with long-term use. This means that your body gets used to the dose, so larger doses are required to get the same pain relief. It is also possible to become physically or psychologically dependent on them. It is very important to have your dose monitored carefully by your GP, and to stay on them for as short a time as possible. However, when taken for pain, most people find it easy to stop taking them once the pain stops. If you think dependency might be a problem for you, you can reduce the dosage very gradually to reduce the risk of withdrawal symptoms.

OTHER MEDICATIONS

Gabapentin
(Neurontin, Gabarone, Fanatrex) This is a drug that was originally developed as an treatment for epilepsy, but more recently it has been used as an effective treatment for pain. It is particularly effective for **neuropathic** pain (nerve pain). It can have significant side effects, though, and some patients cannot tolerate it.

Amitriptyline
(Elavil, Sarotex, Laroxyl) Like Gabapentin, this drug was originally developed for another use, but as knowledge of the drug has increased,

it has become a useful treatment for chronic and neuropathic pain. It is also used as a treatment for depression. For this reason, it can have some stigmatism attached to it. However, the doses that are effective for pain relief are far lower than those that are effective for depression.

Diazepam

(Valium, Antenex, Diastat) This is a group of drugs called benzodiazepines, which are used as sedatives, sleeping tablets, treatment for tension and anxiety, and at lower doses as muscle relaxants. As with opioids, you can develop dependence to these easily, so careful dose monitoring by your GP is essential.

I found it hard to withdraw from my medication (tramadol), and had severe withdrawal, probably because I stopped it too quickly. Before I was diagnosed with hip dysplasia, I took Nurofen Plus (which has codeine in it) when the pain got too severe. After being diagnosed, but before surgery, I took Panadol Osteo every six hours. **Lea-Anne, 40 Australia**

Pre-surgery I was only taking Advil — in very large quantities. It has left me with gastritis (inflammation of the lining of the stomach), unfortunately. **Kris, 47 USA**

I tried a few things, but not much worked really. I think background pain is a bit easier to numb, but pain of moving an essentially injured joint is a bit more of a problem. I always think of how much medication is needed when the paramedics first try to move someone with a limb injury — it's usually a lot!

A lot of medication either made me feel a bit sick/drunk (codeine), or horrible (tramadol). I did, however, love diclofenac. It made my joints feel much smoother and easier to move, without actually killing pain, or making me feel drugged. Like all meds though, not a long term solution! People vary so much though — there really isn't one drug that suits all, unfortunately. **Freja, 45 UK**

Keeping a symptom diary

It can sometimes be hard to objectively assess how much pain you're in, and how disabled you are, especially when you've become accustomed to living with chronic pain. In this case, it can be very helpful to keep a symptom diary. This can be as basic as a notebook or pocket diary, where you give your hip pain a rating out of 10 each day, where 10/10 is extremely excruciating pain that stops you from moving, 5/10 is very distressing pain, and 0/10 is no pain at all. This can really help you see what your situation is.

You can also use your symptom diary as a medical journal (more on this in Chapter 10), to keep a track of medical questions as they occur to you. That way, you can have your questions ready to ask when you see your doctor, physiotherapist, or surgeon.

You might want to track other things like your activities that day, what medications you're taking, other treatments (e.g. ice packs), and when you take medications. This can help you to objectively see what is helping or not, and what impact various activities have on your functioning. You can, of course, add in other symptoms that you want to track (headaches, back ache, swollen knees, and so on) — this is a very useful tool, no matter what your health condition is, and your doctor will love you for the amount of information you'll be able to provide them!

We have designed a DDH pain and symptom record sheet for you, it is available as a PDF on our web site <sutherland-studios.com.au>.

One thing I found useful was to log how my hip felt each day, giving it a score. Without doing this, I found it was too easy to convince myself that my hip pain was no issue and wasn't bothering me. Writing it down and reading it back, it became clear exactly how much of an impact it was having on my life. Obviously, we all have different levels of pain, if any, but it may help you justify your decision whether to have surgery or not. **Dawn, 44 UK**

Non-pharmacological pain relief

There are several other options that you can try that may help with pain. Some of these have some real actions on tissues of the body; others may just help you relax or feel better. But if they work, and that means the pain affects you less, we figure it doesn't really matter how it happens!

ICE

Ice can help reduce inflammation. The hip joint is a deep joint, so whether the cold can actually penetrate right to the joint itself is debatable. But it can help soothe **soft tissues** (e.g. muscles and tendons), and reduce swelling. The trochanteric **bursa** (fluid-filled sac) is much closer to the skin, so if you are suffering with trochanteric **bursitis**, ice can really help.

It's important to only keep the ice on for 20 minutes in a session, no more than three times a day. Put a dripping wet (not just damp) cloth between the ice and your skin. *Never* put ice or an ice pack directly on your skin.

The wet towel is important. An ice pack directly from the freezer can be much colder than 0°C. Skin freezes at 1.8°C. If you put the ice or pack directly onto your skin, it will damage your skin. Using a wet towel between the ice pack and your skin makes all the difference. This sets up a layer of water, which both conducts the cold more efficiently and effectively to your sore hip, and protects the skin from freezing.

HEAT

Heat packs, a warm bath, sauna, or steam room — there are plenty of ways of getting extra warmth. This can help relax muscles, which can help with pain, or sometimes just make it easier to get your hips moving. It can also be very comforting, especially in the winter. Adding Epsom salts (magnesium sulfate) to a hot bath helps reduce 'prune skin' and eases inflammation.

You might like to try alternating heat and ice packs, one after the other, to see if they work more effectively for you in combination.

MASSAGE

Massages can help your muscles relax. If you enjoy them, they can make you feel relaxed all over, which helps to improve your general mood. It's really important to tell the masseuse about your hips, and identify if there are any movements you do not want to do, or areas you do not want massaged. They might be able to work on overworked, tight, or tired muscles before or after surgery. It is best to check with your physio that it is okay, though.

DRY NEEDLING

More physiotherapists / physical therapists now have dry needling in their repertoire of skills. It might also be referred to simply as acupuncture. Dry needling involves having extremely fine needles placed at key points in your body. The needles do not hurt as they are so fine, although it can feel a little odd and unpleasant. It can be particularly useful with tight muscles, especially if there are 'trigger points' in them. The theory is that trigger points are focal, hyper-irritable points within a muscle, where the muscle fibres are very tight; they are tender to press on, and can also be a common cause of pain felt elsewhere in a muscle (referred pain). Inserting the needle is said to release this tension. The doctors at the International Hip Dysplasia Institute, however, say that acupuncture has not been shown to relieve hip pain in controlled studies, although it can have a placebo effect.

CORTISONE INJECTIONS

Steroids in the context of DDH are not the type that weight lifters use, (those are *anabolic* steroids), but *corticosteroids*, often called *cortisone*. These steroids have an anti-inflammatory action similar to ibuprofen and diclofenac (NSAIDs). For some reason we don't call steroids "SAIDs", though, we don't know why!

They are not usually given in tablet form to help with severe pain as, unlike with NSAIDs, *oral* steroids have significant and serious side effects (including weight gain, high blood pressure, diabetes, bone thinning, liver damage, and glaucoma), which can be detrimental, especially when taken long-term (over many months).

So, what areas can be injected with cortisone? With regards to hips, there a few areas that can benefit. They include: the hip joint itself, surrounding tendons (psoas, adductor, and hamstring insertions are the most common), and bursas.

*NB: **You must never have a steroid injection into an artificial joint!***

When steroids are injected in small doses, directly to the spot where the inflammation is present, the risk of these side effects is drastically reduced. These injections are usually given with a small amount of local anaesthetic. This has two benefits: firstly, it numbs the area so that when the steroid is given it shouldn't hurt; and secondly, it can help with diagnosis.

Despite all their training, and the wide range of scans available to doctors, sometimes it's not clear *exactly* what is causing the problem. For example, if your doctor suspects the hip joint is causing the problem, but the presentation of the symptoms isn't classical, they can inject the hip with local anaesthetic — if your pain disappears either in part or total, it tells them that the joint *is* the source of the pain. If the pain *partly* disappears, it can indicate that the joint is only *part* of the problem. If the injection doesn't help at all, it's unlikely to be the joint that is causing the problem. This is not an exact science however, as if, for some reason, the anaesthetic doesn't reach inside the joint, it will still hurt.

X-ray guidance is usually used if the injection site is deep (hip joint, iliopsoas bursa, or tendon). This means a dye that shows up on 'real time' x-rays (fluoroscopy) is also injected, to make sure when the steroid is injected, it is going into the right place. Ultrasound can also be used to guide injections. This is not necessary for all steroid injections, though, especially where the site is close to the skin, like with the trochanteric bursa, as it is easy to directly feel the area that is sore.

Steroid injections are usually done while the patient is awake, however in some circumstances they are done under a general anaesthetic.

Absolute Contraindications (definitely not allowed)

These include: infection around the area or skin, previous **anaphylaxis** (severe allergic reaction), allergy to steroids or local anaesthetic, pregnancy, if you already have a **prosthesis** (direct joint injection not allowed), or general infection.

Relative Contraindications (may be allowed sometimes)

These include: diabetes (steroids, even when given locally, can affect blood sugar control), bleeding conditions (including drug-induced from medication like warfarin), needle phobias, and if it's six weeks to six months before having joint replacement.

Will it Help?

Some people get a great deal of benefit from a steroid injection, others benefit for a period of time, and for some it doesn't work at all[13].

Unfortunately, it is difficult to know which group you're going to fall into before actually having the injection. With regards to DDH, an injection into the hip joint is purely for symptomatic purposes. It can help reduce inflammation and pain, particularly if it has flared up, but it is not going to solve the underlying mechanical problem. It is still worth considering, as the side effects are minimal. You may find you get relief, even if only temporarily, and it is unlikely to do any harm.

Side Effects of Steroid Injections

This list is not definitive; we have just included the more common or potentially serious ones:

◊ **Joint infection** (1 in 17-77,000)[11]

◊ **Anaphylaxis** (extremely rare)

◊ **Post-injection flare** (2-10 in 100)[12]

A flare is when the injected area becomes *more* painful for 48–72 hours after the steroid injection, before it starts to improve. The chance of a flare increases if the area is not completely rested for the first 72 hours. It is not serious, though (just painful). Ice and pain killers can help, and it will settle down. The practitioner giving you the injection should warn you about this possible side effect.

◊ **Depigmentation** (loss of skin colour) (incidence less than 1%)

This will gradually resolve, but may take weeks to months. This side effect is more common in superficial injections, and in people with darker skin.

◊ **Bleeding**. If it's excessive, go to the emergency department.

◊ **Fainting** (usually related to the needle rather than what is injected!)

◊ **Poor blood sugar control** for diabetics.

IONTOPHORESIS

Iontophoresis is a method of delivering medication such as cortisone into the body without injection, using an electric current to painlessly drive the molecules of the medication through the skin. There is more information on this **transdermal** technique in Chapter 5.

HYALURONIC ACID

Also called Hyalgan® or Synvisc®. Hyaluronic acid is a substance that is thought to increase the cushioning effect of the joint fluid, and to increase joint lubrication. These two effects have a pain-killing action. It has been shown to have some benefit in pain reduction and improved functional outcome in **osteoarthritis** of the knee. It can also be used in the hip[13]. treatment involves a course of four to six injections, one per week, which can be repeated, though not in the same six month period.

We hope that the strategies covered in this chapter can help you to cope with your hip dysplasia symptoms. These might be enough to keep you going for quite some time. There are more things you might like to try in Chapter 5, too.

Chapter 4
Surgical Treatments

Hip dysplasia is primarily a mechanical problem, and therefore requires a mechanical solution. This is where surgery comes in. This chapter is only a basic overview. How an individual surgeon does a procedure can vary quite a lot with regards to incisions, technical aspects during the procedure, and the protocols they recommend pre- and post-operatively.

This can seem quite bewildering, as you may think that surely if the aim is the same, how can they vary so much? Surgery as a speciality is extremely adaptable and variable. Techniques vary and improve over time, and there are always advances being made in equipment and information. As a surgeon becomes more experienced in the procedure, he or she may change or adapt things themselves, too.

When we talk about side effects and risks of procedures, we have only included the more common side effects, or the more significant ones. For most procedures, the list of possible complications would be endless; most complication are so rare that they are almost never seen. It would be completely overwhelming to bombard you with every single little thing that could possibly happen, and in doing so you might not appreciate the important ones that you really *do* need to be aware of.

Anaesthetics are covered in detail in Chapter 13, but generally speaking, most operations on the hips are done under a general anaesthetic, where you are put to sleep for the procedure.

Day surgery vs. inpatient stays

For some procedures it is not necessary to stay overnight. In this case, you may find yourself in the day surgery (outpatient) unit. These are specially designed places where you can have your procedure and go home the same day. Some types of steroid injections, hip arthroscopies, and removal of metal work (such as PAO screws), are examples of procedures that can be done as a day case. To be considered for day surgery, you need to be in good overall health, and have someone who can take you home and stay with you for the first 24 hours.

Hip arthroscopy

Arthroscopy is typically associated with knee or shoulder problems. It is only in recent years that the equipment, skills, and experience have been developed to make hip arthroscopy possible.

Arthroscopic surgery is also called **keyhole** surgery. Instead of using a big incision to directly look into the hip joint, it uses several (two to five) much smaller ones (1–3 cm / ⅓ –1" long), and special cameras with long instruments (see Figure 4.1). The images are projected onto a TV screen so the surgeon can see what they are doing. Even though there are more incisions, as they are much smaller, the overall damage to the muscles and other soft tissues is greatly reduced. Because of this, there is less pain, and your recovery is quick. The operation is done under general anaesthetic with **traction** applied to the hip joint. Traction means that your leg is pulled firmly to widen the joint space. You won't be aware of the traction, as you'll be under anaesthetic. The traction widens the joint space, making it easier for the surgeon to see what's going on inside the joint.

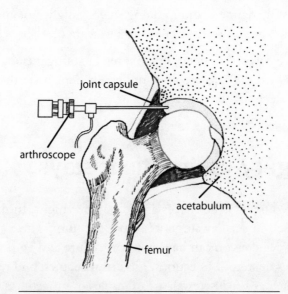

Figure 4.1: Hip arthroscopy.

There are still some problems that can occur inside the hip joint, which are not detected with modern imaging techniques like MRIs and CT scans[14]. An arthroscopy allows the surgeon to directly see the inside of the hip joint. They can make an assessment of the quality of the cartilage, whether any arthritis has developed, and, if so, how severe it is. This can help with decision making for other procedures, and can also indicate how successful these other procedures might be. It can also help establish if DDH is the likely cause of the pain, and assess the quality of the acetabular labrum (the rim of cartilage that surrounds the hip joint).

As we mentioned in Chapter 1, patients with hip dysplasia usually have a hypertrophic (overdeveloped) labrum. This can become damaged due to the abnormal joint stresses on it[15,16].

Hip arthroscopy in patients with dysplasia, or other morphological abnormalities of the hip, does *not* address the underlying biomechanical problem. It has been shown that in the *short* term, arthroscopy can improve pain and function, but in the *long* term this is not the case, with most patients requiring further procedures[2]. It has been suggested that **excision** or **debridement** (removal of damaged tissue) of a damaged labrum in the dysplastic hip could potentially make the hip joint more unstable. This could happen as the already shallow acetabulum is made even *more* shallow, by removing the damaged labrum. In some cases it can lead to the progression of arthritis[2, 17]. In other words, the enlarged labrum offers a certain amount of support and stability to the hip joint, to compensate for the shallow hip socket. If the labrum is trimmed down to a smaller size, this can actually cause the joint to become *less* stable, because the ball at the top of the thigh bone becomes looser within the hip socket.

Hip arthroscopy can help with some conditions, including; torn labrum, loose bodies, osteoarthritis (usually only providing short term relief), **ligamentum teres** injury (ligament connecting the head of the femur to the acetabulum), **femoral acetabular impingement** (FAI), and tissue biopsy.

So, what does this mean for you? Yes, there is a role for arthroscopy in diagnosis and assessment of the dysplastic hip, and it can provide relief in the short term. However, it cannot improve the bone structure of your hip joint, which is the underlying problem that is causing the hip joint to wear out too fast.

RISKS OF ARTHROSCOPY

There are a few risks from hip arthroscopy, including numbness around the incision sites, and numbness in the perineal area. The **perineum** is the area at the top of the inside of your thigh between your legs. The reason that this area can become numb is due to the traction that is applied to the hip joint, which is necessary in order to gain access to the joint. If numbness does occur, it is rarely permanent, and usually settles within one to two weeks[7].

The length of time for rehabilitation after arthroscopy depends on several factors, most importantly whether it is just done for diagnostic purposes, or if an additional procedure is done (e.g. labral repair, or FAI debridement). It may take from six weeks, to several months after more substantial procedures like labral tear repair.

You will usually be on crutches afterwards. If the arthroscopy is diagnostic, this may only be for a few days until you feel comfortable walking unaided. If a larger procedure is done then, depending on your surgeon, you may be asked to be **non-weight-bearing** on crutches for a period of time. The whole recovery period and outcome is variable, depending on what was done and what the underlying conditions are. Be guided by your surgeon.

I had arthroscopy on my right hip prior to a PAO, but it wasn't really my surgeon's first choice. He recommended a PAO, but I'd had an arthroscopy on my left hip before, and wanted to try that for my right hip, before doing something more major. **Louise, 45 USA**

After I had an MRI, it became apparent I had a labral tear and some cysts in my acetabulum. I knew that I would need some kind of joint replacement eventually. My surgeon suggested an arthroscopy first, as it was much less invasive than a replacement. He made it clear that he only regarded it as a temporary measure for my hip, putting off the inevitable, but also that it only had a 50% chance of doing me any lasting good. I felt he had given me a realistic set of expectations, so I went for it, knowing that it most likely wouldn't work, but that it was worth a try. As it turned out, the arthroscopy made my hip joint feel less stiff for about three months, then it started to deteriorate. By four-and-a-half months I knew it had not worked. **Freja, 45 UK**

I had arthroscopy for a torn labrum about 10 years ago, and am about to have that hip replaced. That repair helped my pain for several years, but it has steadily gotten worse over the last five years. **Katie, 47 USA**

Osteotomies

An osteotomy basically means intentionally breaking a bone for the purposes of treatment. These can be carried out on the pelvis side of the hip joint, or on the femur side, depending where the greater deformity is. The aims of both types are to re-establish more normal hip biomechanics and preserve longevity of the hip[18]. In adults with DDH, it is more common that the abnormality is on the acetabular side (i.e. in the pelvis). However, if there is severe deformity of both the pelvis *and* femur, both types of osteotomy can be performed. The outcome is dependent on the degree of arthritis that is present before the surgery. The worse the arthritis, usually the poorer the result[18]. This is true for all types of hip joint preservation surgery. Figure 4.2 shows some places where osteotomies can be made.

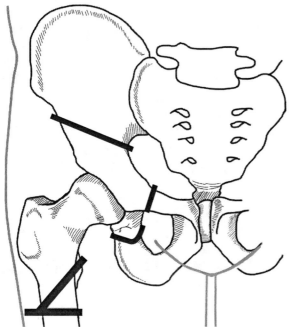

Figure 4.2 Periacetabular osteotomy (PAO) and femoral osteotomy (FO). The cuts marked here are only examples — there are many types of PAOs and FOs, with different positions for the breaks made to the bones.

47

Periacetabular osteotomy (PAO)

Some of the different names for this procedure include: periacetabular osteotomy (PAO), Chiari osteotomy, Ganz osteotomy, triple pelvic osteotomy, shelf osteotomy, and Bernese periacetabular osteotomy. Each of these procedures has minor differences, but the basics are the same. In this book we use the term PAO. *Peri* means "around/about", and *acetabular* means "relating to the acetabulum" (hip socket), so *periacetabular osteotomy* roughly translates as "a procedure cutting the bones around the hip socket".

There are several different types of pelvic osteotomies, however they all do a similar thing. They aim to re-orientate the acetabulum (socket) of the hip joint to make the socket deeper, and make the biomechanics of the joint closer to normal. This can reduce the abnormal stresses that the dysplastic joint is under, and reduce the risk of development of secondary arthritis. If arthritis has already developed, a PAO can slow the progression[19]. Cuts are made around the bone socket to free it, and then the socket is tilted and rotated a certain amount to make better coverage of the femoral head.

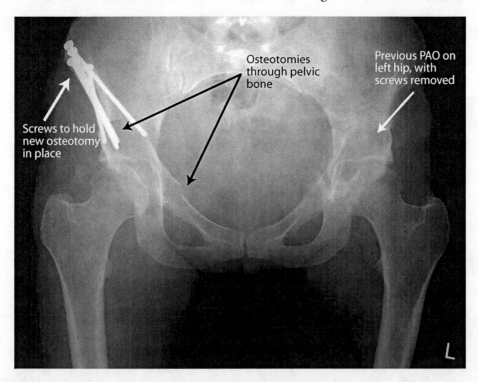

Figure 4.3: X-ray of a female pelvis showing an older healed PAO on the left hip, and a newer PAO with screws still in place on the right hip.

Screws are used to hold the bones in their new positions, while the bone heals (Figure 4.3).

It is important to be aware that the level of arthritis present before the PAO will have some impact on the recovery, and the level or type of activity that can be achieved post-PAO.

Sometimes the arthritis is already too advanced in the hip, in which case undergoing such major surgery is not recommended, as it is trying to preserve a joint that is already beyond help. In this case, some type of hip **resurfacing** or replacement may be your only option. Most importantly, from your perspective, a PAO should improve your quality of life, which is the ultimate goal! Studies have shown that patients can have significant improvement in physical function and pain following PAO[20]. In a review of studies looking at PAOs, a clinically good or excellent outcome was reported in 62–97% of hips, as measured by hip scoring systems[19].

Excellent clinical results were mainly achieved in patients with minimal or no degenerative changes preoperatively[13]. It is well established that the *worse* the arthritis at time of PAO, the *less* likely a good outcome is[22, 23]. Millis et al. studied the longevity of PAOs. The study found that at five years after surgery, 24% of PAO patients aged over 40 (at the time of their PAO surgery) had required a THR. In those with mild or no arthritis a PAO can still produce a good outcome in the majority of cases[24, 25]. Other studies have found that 0–17% of PAO patients under 30 require a THR within five years. So ultimately, if you have no, or minimal, arthritis at the time of being offered a PAO, then it is most likely you will have a good outcome from this major surgery.

A PAO is the most common type of hip preservation surgery (i.e. preserving your own bone, as opposed to a hip replacement), and is a major operation to undergo. It involves several cuts (osteotomies) around the pelvis bones (see Figure 4.2). Some of the muscles are usually cut in order to get to the bone. You may be thinking "Wow, that's got to hurt," and we would be lying if we told you it was going to be pain-free — but there are lots of excellent pain-relieving medications that they give you afterwards which help a lot, so try not let that worry you *too* much. But it is not an easy operation to go through. And never fear, your muscles repair themselves in time, and the bones grow back together.

The exact incision made varies from surgeon to surgeon. Some incisions are along the groin crease, others along the bony bit of the pelvis you can feel at the top of your hip. You will usually need to be in hospital for between three to ten days, depending on your overall health, and how well you manage following the surgery. Sometimes the availability of other medical staff, like physiotherapists or occupational therapists, to make sure you are safe following the procedure, can affect how long your stay is, too.

If you have had muscles cut in your leg, then you may have restrictions like 'no straight leg lifts for six weeks,' and most surgeons require you to be non-weight-bearing (no weight at all can be placed on the operated leg), or just **toe-touch weight-bearing** (toes can touch the ground, but just for balance) for around six weeks.

However, different surgeons have different rules, so it is really important to do what you have been told by your surgeon. The differences usually arise from slightly different surgical techniques (to achieve the same outcome), the fact each hip and person is slightly different, and surgical experience.

Figure 4.3 shows an x-ray of a PAO on the left hip that has had the screws removed but is healed, and a new PAO on the other side which still has screws in place. The arrows point to some of the bone cuts, and the screws to hold the bone in place can be clearly seen.

PAO SCREW REMOVAL

The position of screws in a PAO is quite variable due to different surgical techniques. Everyone has slight variations in anatomy, too. In many cases, they are placed along the iliac crest (see Figure 2.1 on pg 15). You may not be able to feel the screws at all; if you can feel them, they may not bother you.Sometimes, however, you may be aware of them, as this part of the pelvis can be quite close to the skin, and they may rub on tight clothes or belts. If this is the case, then you can ask your surgeon to take them out. You need to wait for all the bones to have healed first, so you need to wait at least six months post-PAO.

Having the screws removed is a very simple procedure, and usually done as day surgery. There are no weight-bearing restrictions afterwards, but you may be a little sore for a week or so. The screws are removed through the same incision as your PAO, so no new scars either!

Allowing the screws to remain in your pelvis will not cause any long term problems; though, if you then go on to need a THR, the surgeon may elect to remove them prior to that procedure. Some people find they just don't like the idea of metal staying in the body, and chose to have them removed for that reason. If the area around the screws becomes infected (which is very unusual), then they would need to be removed.

I had my three screws removed seven years after my PAO, because they had moved and were causing problems. It was a day procedure so I literally walked into hospital mid-morning, and walked out mid-afternoon feeling exactly the same. No post-op pain, no crutches, no medication, barely even discomfort from the incision — it was that simple. **Sandra, 39 Australia**

I had a PAO for my right hip at age 13, and one on my left hip at age 27. I had screws on both hips, but they only removed screws from my right hip a year after the operation. I don't recall that the screws were particularly bothering me, though. With any surgery there is a recovery period, and I was back on crutches for about six weeks or so. The screws in my left hip are not troubling me at all. **Cathy, 39 USA**

I ended up getting the screws out ten months after my right PAO. I had continuing pain for a couple of months prior, with a deep pain (which the surgeon suggested was caused by the diagonal pin), as well as pain on the hip bone (caused by the top of both the other two pins). The pain I had prior to having the pins out was gone, once they were removed. **Chrissie, 31 Australia**

My screws were removed seven months after my left PAO, as scheduled. They were bothering me terribly, and I could feel the heads of the screws on the surface of my skin. I was awake during the procedure, but sedated. I was uncomfortable while the screws were removed, but never in severe pain. I could feel some tugging and pressure, and anytime I felt more than that, they increased my sedation and local anesthetic. I was *so* very pleased to have them removed. I kept them as a souvenir! **Alison, 38 USA**

The process was smooth, and I was happy to get them out! I was told I would need crutches after the procedure but I didn't — the only problem I had was my nerve went numb in the area around the scar. I was told it was a very common side effect, and it is an exception that it didn't happen after my PAO. It's related to a nerve that needs to be moved during the procedure. So now, a year after the screw removal, it is still a bit numb, even though it's slowly getting better with time. **Arpine, 28 USA**

I had the screws from my PAO removed about seven months after surgery, because they were starting to back out on their own, and were causing pain to the soft tissues over them. After all of the medication wore off the next day, I hurt badly enough that I couldn't walk without a stick. I probably should have used crutches for a couple of days, but after using them for three months following the PAO, I just couldn't bring myself to use them again.
Louise, 45 USA

The screw removal was very simple and painless. The doc did both PAO sides at the same time, and I left walking under my own power without crutches. It was an outpatient surgery, but I was knockied out for the procedure. The screws weren't bothering me, but I wanted them removed so I could have MRIs in the future, if my hips began to bother me again. **Shelley, 37 USA**

PAO COMPLICATIONS

What is interesting about PAO surgery is that the surgical experience of the surgeon is very important. You may say "Surely this is true for *all* types of procedures?", and of course it is. However, while hip replacements and arthroscopies are carried out by a large number of **orthopaedic surgeons**, the number of surgeons doing PAOs on a regular basis is *much* fewer. The rate and incidence of complications after PAO has been directly linked to surgical experience. A review of studies looking at PAOs acknowledged a significant learning curve associated with this particular procedure. This suggests that the complication rate may diminish with increased surgical experience[19, 21, 26-29].

The complications that can arise from PAO include:

◊ **Sciatic or femoral nerve damage**: The sciatic nerve runs behind the leg, and close to where some of the bone cuts are made. It can occasionally be damaged; the risk of this is reported between 1-4% [19, 21, 29]. This can result in foot drop and numbness in the leg, which usually resolves with time, but can occasionally be permanent. You can also get neuropathic (nerve) pain and muscle weakness, if the part of the nerve working the muscles is affected.

◊ **Lateral cutaneous nerve damage**: This is a tiny nerve that can get cut around the incision site. It provides sensation to the outer region of the thigh, sometimes extending down to the knee. It does not supply any muscles. This has been reported between 2-33%, and can result in numbness to the thigh[19, 21]. Sometimes abnormal sensations like tingling and pain can occur.

◊ **Bleeding**: As a PAO is major surgery, and the pelvic bones have an excellent blood supply, there is a significant risk of bleeding during surgery. Very occasionally some major vessels may also be damaged. The risks of requiring an **allograft** blood transfusion (not your own blood) during a PAO are 20%. The risk drops if the pre-operative **haemoglobin** is more than 12 g/dl, and it is possible to not need any extra blood at all[30, 31].

◊ **Nonunion**: This is when the cut bone ends to do not heal back together. It is a risk when *any* bone fracture occurs, whether it's intentional or not. Luckily, the incidence of this complication after a PAO is very low, with rates of around 1% reported[19].

◊ **Infection and deep vein thrombosis (DVT)**: These are discussed in detail below, in the second on general complications from surgery. The reported rates for DVT after PAO are 2–5%, and for infection are 2% [7, 19, 27, 32].

I had a pelvic stress fracture after my PAO. It developed about three months after the operation, when I was off crutches and feeling great — so it was a big disappointment. I was very demotivated and down when I found out about it. As a result, I was limited in the exercises I could do (so I gained weight), and had to wear a bone stimulator, which after three months I started ignoring a bit, since it was very hot and the stimulator wouldn't stick to my skin. The fracture is not completely healed yet, but it looks like it's getting there, and I hope for the best! It is not bothering me much, luckily, but still I would prefer to not have it.
Arpine, 28 USA

I had a bleed of a branch of the hypogastric artery after my first PAO, which was treated with coil embolization at four days post-op. They couldn't figure out why my hematocrit and hemoglobin levels kept dropping after surgery. I ended up needing 8 units of blood and 4 units of **platelets** in all. They think it might have been a drill that nicked the artery during surgery?

I was anemic upon discharge, and needed to go regularly for bloodwork until my levels were almost normal. I was very weak and tired for the first month, and the iron supplements gave me awful constipation. MiraLAX® was my best friend.

After my second PAO, I had a nonunion of the pubic osteotomy, diagnosed at my 12 week check. Exercise was limited to pool only, then later, swimming and finally, stationary/spin bike. I used an ultrasound bone stimulator for 20 mins/day for months. 18 months later, I'm told it is finally healed. *Yay!* That side feels great now.

Finally, I slipped in the shower a year ago, and injured my right hip (first PAO side, which had completely healed and felt great). MRI shows a torn labrum. However, because my symptoms are not typical of a labral tear, the hip arthroscopy specialist can't tell me if the labral tear is the cause of my pain, or if arthroscopy will help. I'm still using a cane off and on during the day. Since it's been a year, I've decided to schedule the arthroscopy. **Lara, 40 USA**

PAO REHABILIATION

As you have probably guessed, rehabilitation after a PAO is a long haul. Your muscles, tendons and bones all have to heal, and not only that, but in slightly new positions to those they've been in for your whole life. You will be on crutches for between six to 12 weeks, and once you are over the 'non-weight-bearing' period, there will be lots of exercises to do, to strengthen muscles that will have wasted or 'gone to sleep' during this time.

You will definitely need a physiotherapy / physical therapy program, and hydrotherapy is usually also recommended. This can be particularly useful while you are on crutches, as it allows you to move your hip while the water supports you, which believe us, feels wonderful! Most people find they can return to most activities by 12 weeks and impact sport by six months [7].

Femoral osteotomy

Femoral osteotomies are a less common surgery associated with DDH. They are necessary if the major deformity is on the femur side. Femoral deformities are generally either **coxa valga** (where the angle the femur makes with the head is too wide), or **coxa vara** (where the angle is too small). See Figure 4.4 below. Occasionally there is a rotational deformity of the femur, in which case a de-rotation osteotomy is done.

The scar from a femoral osteotomy is usually 10–20 cm / 4–8" long, and on the outside of the thigh. It is usually necessary to cut some of the muscles in order to access the hip, which means that lots of strengthening exercises are needed after the surgery, to get the muscles working properly again.

You will probably be non-weight-bearing (on crutches) for a period of time; this is usually six to eight weeks, depending on your surgeon, and the exact nature of the procedure. There may also be restrictions on movements, depending on whether any muscles or tendons were cut. The rehabilitation process is generally quite long, requiring lots of physiotherapy and/or hydrotherapy. You will probably find you can return to most activities by three months, and impact sports by six months. However, each case is different, so be guided by your surgeon and physiotherapist.

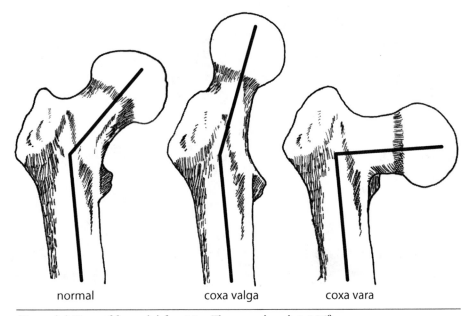

normal   coxa valga   coxa vara

Figure 4.4: Types of femoral deformities. The normal angle is 125°.

Hip resurfacing

Hip resurfacing is a type of joint replacement that aims to preserve more of your own bone on the side of the femur, compared to a total hip replacement (where the top of the femur is completely removed). Resurfacing has recently been accepted throughout the world as an ideal option for younger patients, due to the preservation of bone, better stability (due to the larger head used), and improvement of biomechanics[33]. It has also been shown that a better hip range of movement can be achieved with a resurfacing over a THR[34].

In Figure 4.5 below, the image on the left shows a resurfacing; only the surface of the head of the femur has been replaced. On the right is a THR, and you can see the prosthesis is completely replacing the whole femoral head, and extends far down into the femur. The acetabulum (socket) part of the joint is replaced in both procedures.

There is not a lot of information available on resurfacing in patients with DDH. There are several reasons for this. One is that you need relatively normal anatomy (apart from the arthritis), particularly on the side of the femur, which can be hard to find in dysplasia patients. Also, in younger female patients, there is a theoretical risk that the wear of a metal-on-metal prosthesis could possibly produce microscopic metal fragments, that might have an impact on pregnancy and childbirth, so a resurfacing may

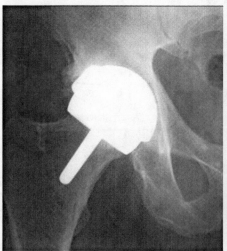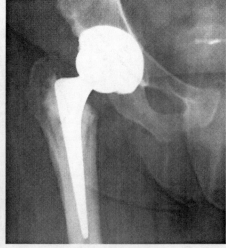

Figure 4.5: The x-ray on the left shows a hip resurfacing, the x-ray on the right shows a total hip replacement on a dysplastic hip. The dense white areas are the prostheses.

be avoided for women who want to have children in the future. For people with a leg length discrepancy, this is more difficult to correct by resurfacing, and sometimes not possible with this procedure.

Studies looking at resurfacing in hip dysplasia have shown that good clinical outcomes can be achieved. However, careful patient selection is necessary, as the procedure is not suitable for everyone, and studies looking at long term outcomes are also needed[33-35].

The actual resurfacing surgery is not drastically different to a THR with regards to incision and rehabilitation. The incision may be longer than a THR incision, as the surgeon needs more room to move around within your hip joint, because the femoral head and neck are not removed.

I have a very long scar — at least 12 inches long — as a resurfacing requires a larger ball head, plus the extra arthritis I had made this a necessity.
Laurie, 50 USA

RESURFACING COMPLICATIONS

These are similar to those seen with a THR, and can include: fracture (around the neck of the femur), loosening of the prosthesis, and the need for **revision** (replacement of the prosthesis). The risk of dislocation is much less than in THRs, because of the considerably larger femoral head component used. Studies into rates of complications in resurfacing for DDH are scarce; one study reported a complication rate ranging from 0–6% [33, 35].

As you can read below, resurfacing is a very popular option for those hippies who are candidates for this surgery.

I chose a resurfacing after a lot of thought and research. For me, it was a no-brainer, I mean, why have more bone removed when you can have less? The fact that resurfacings have a super-low dislocation rate, a great range of motion, and (supposedly) a great lifespan, were perks. I was most concerned about the prospect of repeated revisions as I aged. Since resurfacings are usually much easier to revise than a THR, it made total sense to try to start out with a resurfacing, and revise to a THR, rather than having a THR as my starting point.

57

Sure, resurfacings carry a small risk of failure (as do all joint replacements), but after doing the research, I thought that the benefits outweighed the risks. So far (two and a half years on), it has paid off. I am very happy with my resurfacing — it changed my life. **Freja, 45 UK**

I had a PAO and femoral osteotomy 18 years ago, which were successful. However, 18 years ago, techniques weren't as they are now, and after nine hours in surgery, I was very very poorly, and it took me almost two years to fully recover and go back to work. A very tough time — so I was terrified that I was going to face the same issues this time around.

I think I was in denial about how bad things were, and only sought treatment when I could barely walk. I did all the research about the options, and became quite determined that a resurfacing would be the best option for me, as I didn't want any restrictions, and I am quite sporty. I saw four surgeons — three told me I was not a candidate for a resurfacing, but one said it was a possibility, but I had to prepare myself that I may wake up after surgery with a THR or, because of the severity of my dysplasia, I may wake up with a BHR (Birmingham hip resurfacing), but also have to have a femoral osteotomy. I was really scared going into surgery, but felt I had the best surgeon and this was my best chance. I woke up with a BHR (and no femoral osteotomy!).

I'm over the moon with my resurfacing — I have no restrictions, I am really fit, and I teach group exercise (a kind of play job on top of my full time job). You would never know I had had a problem. I am pain free, I don't even think about it now. I would advise patients to seek out the best surgeons — resurfacing is a very skilled process, and there are a handful of surgeons that do it well, and even fewer who will take on difficult cases like my own. On recovery, find a physio who understands dysplasia and hips. **Tina G, 41 UK**

I was a good candidate because of my age (50). I looked into a PAO and was scheduled for one, but the day before my surgery, my surgeon had to cancel because his father was having emergency cancer surgery the same day, and he felt that his head wouldn't be in my surgery. I was incredibly apprehensive about the surgery to begin with, because of my age, so I took this as a sign from above that I wasn't meant for a PAO. I hadn't felt comfortable with the lack of testing the specialist did, as well as the hospital procedures, so it just seemed like the best idea to cancel it entirely. Plus, every day I waited, I got older and the arthritis got worse, and they couldn't reschedule me for another four months. With the PAO, the doctor told me that I'd probably have to have a THR within five years, so I really questioned why I would put myself through this long recovery time, only to have to do another surgery in five years. It was a very hard decision. I decided to go for a resurfacing instead.

I recommend anyone in the situation of being too old for a PAO, and too young for a THR, to find an orthopaedic specialist who does hip resurfacing, and discuss that as an option. It helps to save bone for a later THR. I feel like I got my life back. As much as I want to use that handicapped sticker I still have, I just can't justify it. I feel too good and I don't want bad karma! As far as my recovery went, I had no problems. I walk great now — no limp. In fact, I recovered too fast — I didn't get to take advantage of my husband and kids enough!
Laurie, 50 USA

Total hip replacement (THR)

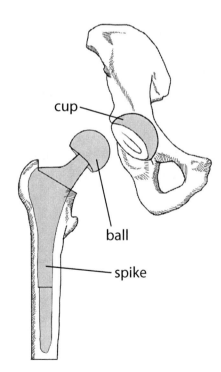

cup

ball

spike

Figure 4.6: Total hip replacement, showing femur and socket prostheses.

Total hip replacement (THR), or full hip replacement (FHR), is one of the most common orthopaedic procedures carried out throughout the developed world. It has been used as a treatment for hip dysplasia, or more specifically the secondary osteoarthritis *caused* by hip dysplasia, since the 1960s. Although orthopaedic hip surgeons will do thousands of these, it is still important to find one who is experienced in joint replacement specifically for DDH. This is because the hip anatomy is different from normal, and many patients have had previous operations on their hips, which can make the procedure trickier. Figures 4.5 and 4.6 show the main features of a hip replacement — the new hip socket (cup, and the new

59

femur head (ball). The spike is firmly inserted into the femur. Different brands of hip prostheses have slightly different shapes.

For those who have undergone joint preservation surgery, such as a PAO, it is possible to have a THR after this procedure. If the same approach is used, a PAO may in fact improve the results of a THR due to the previous restoration of nearer to normal anatomy[36]. It is possible to have a completely dislocated hip, and still undergo a successful replacement, although specialist techniques (and therefore specialist surgeons) are required[37].

There are two main types of hip replacement, cemented (cement used to secure the prosthesis into the bone), and non-cemented (the prosthesis is just fitted very snugly into the bone with a surface that "sticks" to the bone). This can apply to both parts (the ball and socket), or just one part. Non-cemented replacements are usually preferred in younger patients (under 55), and contain special coatings that allow the bone to grow into the prosthesis[7].

There are three main options when it comes to the materials used to make your hip replacement. All of them are **biocompatible** (are not rejected by the body), resistant to wear and corrosion, and meet very high standards. Ceramic (aluminium oxide), metal, and plastic (high-density polyethylene) are the materials used. The prosthesis can be in various combinations (e.g. metal socket and metal femur head, plastic socket and ceramic femur head, and so on). Each material has its pros and cons, and your surgeon will advise you on what they think is best for you.

All prostheses wear down slowly, and create wear particles (see Chapter 15). It is not known how the metal ions that wear off metal prostheses affect an unborn fetus, so if you are of child-bearing age, and wish to have children, your surgeon may advise against a metal prosthesis.

This is a big operation, and will involve being in hospital for several days, possibly even over a week. Both the socket and ball part of the hip joint are replaced. There are many different designs of prosthesis, and the type used depends on several factors including the surgeon's preference, your age, your activity levels, and your anatomy. Consequently, the incisions, and therefore the scars, can vary. Usually a 15–20 cm / 6–8" incision is needed, which is generally over the outer aspect of the thigh. If you have had previous surgeries, though, the surgeon may be able to go through a previous incision.

As your damaged dysplastic joint is entirely removed and replaced, all the pain from the osteoarthritis is taken away immediately. Once you are over the surgery, you will usually see a dramatic and rapid difference in your quality of life. Due to the abnormal anatomy of dysplastic hips it is occasionally necessary to have bone grafts as part of the procedure, or a type of osteotomy on the femoral side to either rotate or shorten the femur.

THR COMPLICATIONS

The following risks usually occur in less than 2% of cases, however higher rates have also been reported[7, 18].

◊ **Femur fracture**: This can occur during the operation when the implant is being put into the femur. If it happens, the problem can be fixed during the operation, the only impact for you may be a change in rehabilitation protocols, and the risk of nonunion.

◊ **Nerve injury**: The sciatic nerve can be damaged during the procedure, either by direct injury or due to stretching (especially if your leg has been lengthened during surgery). Usually this is not permanent, but can take many months to resolve. Symptoms of sciatic nerve injury include muscle weakness, numbness, and pain. The obturator nerve can also be affected.

◊ **Deep vein thrombosis (DVT)**: The incidence of DVT following joint replacement, if no precautions are taken, can be as high as 70%[38]. However, the standard use of preventative blood thinning injections and compression stockings, in addition to pneumatic devices in the early post-operative period, has reduced the incidence to under 8%[39]. In many reports studying patients with THR for severe dysplasia, there are almost no reports of DVT[37, 40, 41].

◊ **Infection**: This is rare and is discussed in detail below, in the section on general complications from surgery.

◊ **Aseptic loosening**: Sometimes the parts of the replacement can get loose. This occurs from wear and tear of the prosthesis over years. It is more common in younger patients, and can be painful. Microscopic wear particles from the prosthesis get into the bone around the prosthesis, and may cause an inflammatory response whereby the bone dissolves around the prosthesis (**osteolysis**). If this occurs, usually a revision of the joint is required[42]. The rate

61

of aseptic loosening is generally around 10% of patients with a first THR after 10 years, and 15% in patients with a revised hip replacement [85,43]. Aseptic loosening is the main reason patients end up needing a revision.

◊ **Dislocation**: This can occur because all the ligaments and lots of the muscles are damaged during the procedure. Until healing occurs, there is very little, apart from the prosthesis, offering stability to the joint. Occasionally the joint can dislocate because the socket part of the replacement is not at the best angle. This is why **hip restrictions** (limited movements) are so important following a hip replacement. You will know if your hip dislocates! Usually there is a pop, *severe* pain and an inability to walk. A dislocation can sometimes just be put back in by the emergency department, but it often requires a general anaesthetic.

There is more in-depth discussion about post-THR hip restrictions and THR complications in Chapter 15.

During my THR surgery, my surgeon lengthened my leg (which was a few centimetres short), and straightened it (it had been broken and rotated outwards during childhood surgeries). That took quite a while— a good year — to adjust to, and I had quite a few problems with pain from the nerves being stretched to the new leg length, my obturator and sciatic nerves in particular weren't very happy for many months. I also had a small **pulmonary embolism**. **Denise, 47 Australia**

Usually you will be encouraged to fully weight-bear straight away after a THR, although you will need the help of crutches, a walking frame, or a stick for some time. If you had additional procedures due to the severity of the anatomical abnormality, such as femoral shortening, it may mean that you are required to be non- or partial weight-bearing for a period of time. Be careful to follow the instructions given to you by your surgeon.

There are two main parts to getting over a hip replacement. The first is the presence of 'hip restrictions,' these are designed to prevent your new hip dislocating while the muscles can recover enough to give your joint the stability it needs. Very occasionally, these restrictions can be in place for life, but for most people they are only for the first six to 12 weeks following the surgery. These are so important we have devoted a whole chapter to them, see Chapter 15. Secondly, as we have mentioned in relation to osteotomies, after a hip replacement you will need to work with a physiotherapist to regain all the strength in your muscles, and achieve a normal walking gait.

Revision hip replacement

Unfortunately, joint replacements don't always last forever. While they can last a life time in the older person who has lower activity levels, in younger active people, they undergo more stresses, and can wear out sooner. The good news is that, although it is a much bigger surgery than a primary hip replacement, it is still possible to have a revision. Aseptic loosening is the most common reason people end up with a revision THR[85].

In a review of studies looking at patients under 50 years of age undergoing hip replacement, the reported revision rates at 10 years were between 64–90%[44-46]. Patients who underwent a total hip replacement with DDH were 1.5 times more likely to need a revision, compared to those without DDH[47].

The revision procedure is more complex than a primary THR, because the bone around the prosthesis can deteriorate in function, as the implant moves around[7]. Additional techniques, such as bone grafts and special prostheses, are usually required. The complications are the same as for a primary THR, however the rates are generally higher[7, 48].

I had a hip revision on my right hip six years after my THR due to osteolysis. It's been a year since my revision, and I continue to experience pain and discomfort. Diagnostic tests and lab work look good, so my pain may be related to a nerve. This recovery period has been the most challenging and painful of them all. It has impacted my ability to work. **Cathy, 39 USA**

You may be heartened by Margaret's experience though!

After a lifetime of hip problems, bilateral dislocation at birth, osteotomies, and finally a careless driver driving into the back of my car, I had my left hip replaced on my 24th birthday by Sir John Charnley himself! Two children later, I had my right hip replaced when I was 40. I was 60 last month and both hips are still going strong — yup, the left is now 36 years old (heading for *Guinness Book of Records* I hope). Without the hip replacements, my life would have been totally different, and I have not a single regret about either. **Margaret, 60 UK**

Deciding when to have surgery

Deciding when to go ahead with treatment or surgery is another challenge. There will be some factors outside your control, that will influence when you can have surgery, such as waiting lists and surgeon availability. One of the luxuries (if there is such a thing) about this condition and its attendant surgeries is, while surgery is usually necessary, it does not need to be done immediately, giving you time to plan the surgery into your life. Many surgeons, even on the NHS in the UK, will be able to give you some flexibility with regards to timing.

SYMPTOMS AND QUALITY OF LIFE

For the majority of people, your quality of life will be the main factor for when you decide to have surgery. If you have been told that the arthritis is too advanced for a PAO, you may feel you can tolerate your symptoms for a time, prolonging the time until your joint needs to be replaced (which is important, as joint replacements do not last forever, especially in more active younger patients). Alternatively, as we have previously discussed, joint salvage surgeries like a PAO do better the *less* advanced the arthritis is, so this can be a reason to take the plunge into surgery sooner.

In some countries, the wait for surgery can be as long as a year or more (for example, if you're going through the public Medicare system in Australia). While your hip may be bearable now, it might *not* be so good in a year — which can be another reason to look at starting the process sooner rather than later. Keep in mind that arthritic joints can deteriorate suddenly, going from bearable to crippling in short order.

You will initially, of course, be in more pain and have more mobility problems *immediately* after the surgery — there's a lot of recovery to do — but with every week you will be so much better, and after a year or so you won't know yourself. Most surgeons and physiotherapists say that it takes a good year after major surgery like a THR or PAO to really be back to normal (and better than what was normal for you before!).

For many, there comes a point when it is just too much to carry on living with the pain and restrictions their hips are putting on their life. This is a very personal decision. It is really important to do the right thing for *yourself*, and not to compare yourself to others. You're the only one who really knows how bad things are, and how much it is impacting on your life physically, socially, and emotionally.

I did not wait ... I was in too much pain. I do have to admit I was afraid and had to mentally prepare for the journey. My body said "Do It". I think I should have started taking Xanax or something for panic and anxiety, but my doctor wouldn't give me anything. **Ina, 41 USA**

I didn't wait for surgery. As soon as the doc gave me the diagnosis and told me what I needed to have done, I scheduled my PAO surgery. Then I took the information home and read it, and became really freaked out by my decision. I think if I would have read the info and learned about the surgery before scheduling, I never would have scheduled it!

I scheduled the second hip when I went for my six week post-op follow up. I wasn't having any pain yet, but on the x-rays the 'uncorrected' hip was actually structurally worse. I knew I would need it done, so I chose to do it preventatively, and am very happy with that decision. My doc thinks I preserved the joint well enough that I might not need to have a THR in the future. **Shelley, 37 USA**

I didn't wait for my first PAO, I booked in the surgery immediately after my DDH diagnosis. As I was using the Australian public health system, I was on the waiting list for over six months, so I had lots of time to read about the condition, research the options available, and basically get used to the idea.

I did wait for my second PAO. I was diagnosed in my early thirties — at a time when I was considering starting a family. I chose children over surgery as I felt I was running out of time to have children due to my age, and also because

my hip was only mildly symptomatic at the time. I guess it was just a case of which was a higher priority. Thankfully I was still a PAO candidate three years later, although I did have some arthritic changes during that time.
Sandra, 39 Australia

I got my diagnosis around four months before I decided to have surgery. By the time I got in to see the orthopedic surgeon, I was in so much pain, my life had been so disrupted by pain, and I felt like I had tried everything else. I scheduled my surgery date when I first met with my orthopedic surgeon.
Meghan, 28 USA

I waited for two years after my diagnosis. The first surgeon I saw really scared me by saying that if I didn't have the PAO in six months, I wouldn't be able to walk properly anymore. He would not answer any of my questions and just acted like I should trust him and not ask him anything.

He even lied on my appointment report by saying that he found some odd features from a CT scan on my hip. I never got a CT scan, though. He was really inexperienced, and I realized that he was trying to get a patient to practice on.

After that experience, I was wary of going to just any hip surgeon. So I took my time doing a lot of background research on everything from physician experience, ratings, board qualifications, malpractice suits, insurance coverage, and reading the literature. I saw three additional hip surgeons who did PAOs before I finally settled with Dr. R. Approval for my appointments with these surgeons took a while to get from my insurance company, and the surgeons were pretty booked, which also delayed things. I'm a graduate student, so I wanted to schedule my surgery for the summer. This also delayed things a bit.
Shilpa, 27 USA

I did not wait to have surgery. After my diagnosis I was referred to Boston to see their 'hip team', and we discussed the options. After a failed arthroscope, I went ahead with the PAO surgery. I was a full-time student at the time of the diagnosis and was unable to navigate a college campus because I was in horrible pain. I was experiencing subluxations and felt unstable. My hip joint was already showing signs of arthritis (at 20 years old), and I didn't want it to get worse. I was told I would need a THR within 2-5 years if I did not correct the dysplasia. It was a fairly easy decision to go ahead with surgery. **Jill, 20 USA**

Work and study

Your work or study commitments are likely to be a factor in your decision making process. You need to carefully weigh up the benefits of surgery, and the timing of recovery, alongside what you do. Your entitlement to sick pay may also be something you need to consider. If you are a student it may be prudent to consider undergoing surgery during the school holidays or taking some time off. If you have important exams coming up, it might be good to get them done and out the way, or put them on hold for a long enough period so that you have time to fully recover from the surgery — you do not want to be trying to study, and having the stress associated with exams, while getting over the operation.

At work, if you have a good working relationship with your boss, openly discussing the timing and expected impact on your functioning with them should hold you in good stead, and consequently may decrease your stress levels (hopefully!). One thing to be aware of is most of the major procedures (THR, PAO, resurfacing) will take at least three to six months to fully get over, and it can take even longer to feel completely back to your normal self. Bear this in mind when planning your surgery. It is also massively dependent on what you do. If your work is a desk job and you can work from home, you'll obviously be able to return to some level of work far sooner than for example someone who is on their feet running round all day! We further cover workplace negotiations in Chapter 6, and returning to work in Chapter 9.

Family issues

You will need to consider your family's needs when considering major surgery. Family members are usually the main carers for hippies, after hip dysplasia surgery. You will need a significant level of assistance for some weeks after surgery, and a lower level of assistance for another few months. The decision therefore needs to be one you make together. We discuss family issues in more detail in Chapter 9.

PREGNANCY AND YOUNG CHILDREN

Pregnancy puts more strain on the hips. This is due to the extra weight, and the physical and hormonal changes that happen in pregnancy. And there is no doubt about it — having major surgery and recovering from it with young children around is never going to be easy. However, there are plenty of people who manage it. Many people find their children are their inspiration during their recovery.

If you know you need the surgery, and want to start a family, you need to think carefully about what to do first, taking into account your current symptoms. There is the potential for them to get worse (remember we say *potential*, as although likely, it is *not* a given that your hips will get worse during pregnancy). There are also numerous other factors in your life to consider when starting a family. All the hip surgeries on offer for DDH aim to improve quality of life. For the younger population of hippies, this also means your sex life and pregnancy. While male DDH patients never go through pregnancy, their partners will have to shoulder most of the childcaring duties while they are recovering, so this needs to be taken into account too.

You will be able to have both sex (phew — sigh of relief) and (if you are female) a successful pregnancy following any surgery. Sex is usually not something most people feel able to ask their surgeons about, and quite honestly is something many surgeons may not comfortable talking about. We discuss this further in Chapter 16.

I delayed having a third baby because I was given about a year before I was no longer a candidate for PAO, so I didn't want to have the baby then not be able to breastfeed for as long as I liked, or to have family look after a baby plus me and my 2 and 4 year olds! **Lea-Anne, 40 Australia**

I put off children when I was younger because of my hips, and now will not have any. **Katie, 47 USA**

I never wanted kids, so this was not an issue for me. However, my surgeon took some convincing that I *really* didn't want kids and I felt just a little insulted — I mean, a bloke wouldn't get asked three times if he *really* didn't want kids, would he? **Freja, 45 UK**

This was one of my major concerns being 22 at the time. I didn't have to delay surgery as I wasn't in a position to start a family then, but I certainly wanted to know if it would restrict my future plans. I could hardly ask a friend of the family (aged 77) whether it hindered her sex life or future family plans! She was the only person I knew who had a THR! And that would have been an interesting if slightly embarrassing conversation to have! **Rhianna, 28 UK**

If you've had childhood hip surgeries, pregnancy can sometimes put a great deal of added strain on your hips, and this can be quite painful. However, many women who've had childhood surgery for dysplasia don't have any problems at all.

My childhood surgeons told my mother that they weren't sure if my (surgically corrected) dysplastic left hip would 'survive' pregnancy. True to form, both my pregnancies were accompanied by a lot of hip pain. I had to have Caesarians for both births as my pelvic outlet was quite deformed (possibly as a result of my childhood surgeries). After my second child was born, I was advised not to have any more children as my hip was unlikely to cope. **Denise, 47 Australia**

For the last two weeks of my first pregnancy I did have some hip pain at night if I slept on my side. Changing position solved the problem. My first daughter turned out to be 10 lbs, so I think that's why. With my second daughter (8 lbs, 9 oz) I had no pain or problems whatsoever. **Betsy, 48 USA**

Complications from surgery

All surgery has associated risks. Most are rare, but you need to be aware of them, so that you can make an informed decision. We have discussed the risks specific to each procedure above, however, there are a few risks that are found with *all* types of **orthopaedic surgery**, and these are discussed below. The figures we give are only a guideline; you need to ask your surgeon for their exact figures for your hospital and procedure. They are: bleeding, blood clots, chest infections, nerve damage, urinary tract infections, and wound infections.

BLEEDING

It sounds obvious, but bleeding is an important complication to think about and appreciate. Steps are taken during any operation to reduce blood loss, but with the bigger procedures, blood loss can be significant. This can mean you become anaemic after the operation.

For a man, a normal haemoglobin is more than 13.0 g/dl, and for a woman, over 12.0 g/dl. However, a transfusion is not usually required unless levels drop below 8g/dl. Above this level, the low levels are usually well tolerated. Procedures for blood transfusions vary drastically from country to country. For example, in the USA and Australia you can be asked to donate your own blood (**autologous** donation) in the weeks running up to surgery, so that it can be given back to you if you need it after surgery. Family or friends with the same blood type can also donate to you (**directed/ designated donation**). This is not an option in the UK, where all blood donations are anonymous. Some surgeons use a **cell saver** which is a special machine that recycles your own blood that you lose during the operation, and gives it straight back to you immediately during surgery. This avoids any of the risks associated with receiving a donation from the blood bank/ anonymous donor.

In the hospital, I had to have two pints of blood transfused due to extra blood loss during the surgery. With the additional arthritis he discovered, the surgery took longer so I had more blood loss. I was frustrated about having to use someone else's blood, since I didn't have any taken prior to surgery. **Laurie, 50 USA**

BLOOD CLOTS

When a blood clot forms in the leg it is called a **deep vein thrombosi**s, or DVT. A blood clot in the lungs is called a **pulmonary embolism**, or PE. A PE can occasionally occur from a clot forming in the lungs, but more commonly, it occurs when a bit of clot from a DVT in the leg breaks off, and then travels to the lungs. Risk factors for developing a clot (other than surgery) are immobilisation, pregnancy, previous history of a clot, smoking, obesity, and certain blood conditions.

Symptoms of a DVT

A leg that is swollen (more than 2 cm (~1") greater than the other leg), painful, red, and with dilated veins on the surface; but occasionally there may be no symptoms at all.

Symptoms of a PE

Chest pain, a fast heart rate, feeling breathless, low oxygen levels, collapse, dizziness, and cardiac arrest. A PE is a life threatening problem.

What to do

Blood clots are a very serious complication, and if you even vaguely suspect you might have developed a blood clot, seek emergency medical help. Better to have a false alarm than to miss a real DVT or PE!

Because of the seriousness of this complication, you are made to wear DVT stockings (also called compression or TED stockings) after hip surgery. Not all surgeons get their patients to wear the stockings, however, so follow whatever their advice is.

While you are less mobile or in hospital, you are given blood thinning injections (low molecular weight heparin or Clexane®), to help reduce the risk of these problems developing. Often you will need to continue these injections for a month or so as a preventative measure, giving them to yourself at home. The nurses will show you how to do it; while it's unpleasant, it isn't hard to do.

It is also important to do ankle exercises while you are in bed to keep your calf muscles working. Your nurses and occupational therapist will show you what to do. You can do these frequently (every 5–10 minutes if you remember!), and you can start immediately after surgery.

Two basic ankle exercises which you can do lying in bed are:

◊ Point your toes, and then flex your feet so the toes point to the ceiling (like pumping the accelerator on a car).

◊ Turn your feet towards each other, rotating at the ankle (pigeon-toed), and then rotate them outwards (duck footed).

If you do develop a DVT or PE, the treatment is to thin your blood, which helps to break the clot down, and stops any more clots from forming. This usually involves being given larger doses of the blood thinning injections, and being started **anticoagulant** medication such as warfarin.

Warfarin is taken orally. It takes a few days to get the right levels in your system, so you will need to stay on the blood thinning injections until a good level is reached with your oral medication. Warfarin levels are monitored by measuring your **International Normalised Ratio** (INR). The desired level following a PE/DVT is a range of 2–3, (1 is normal). Once this level is reached, and your warfarin dose is sorted out, the injections can stop (this usually just takes a few days), and you can get out of hospital.

You then need to stay on the warfarin for three to six months, with frequent blood tests to check your INR levels. Warfarin is a tricky medication to be on, as foods and drinks which have vitamin K in them will affect your INR levels. You need to avoid eating large quantities of leafy greens, and avoid things like green tea. Many medications interfere with warfarin too. Your GP will give you guidance on all this.

A couple of days after my THR surgery I developed a pulmonary embolism. It was only small, thankfully. The nurses noticed that my oxygen levels were dropping, and I was a bit breathless. I didn't feel it very much, though, as recovering from the hip surgery was a more painful and immediate problem for me! I had a chest x-ray and a CT scan, which confirmed the existence of the PE. I was given higher dose Clexane injections, and had to have extra oxygen for a few days (just a little tube that sat under my nose). The worst thing about all this was having to be moved on and off beds and trolleys to have the scans, it was excruciatingly painful to my hip. I had to stay in hospital for a few extra days while they got my warfarin dose right. I was on warfarin for six months, and had to wear DVT stockings for a few months. **Denise, 47 Australia**

Ten weeks after my most recent hip surgery, I twisted my ankle. Later that day, I felt pain in my calf, and ascribed it to a sprain or muscle strain. I waited a week before going to the Emergency Room, and by that time my calf was swollen, red, and warm.

The ER physician sent me for a special kind of ultrasound scan, which confirmed his suspicion — a DVT behind my right (non-operated) knee. It took three days in the hospital for my clotting time to rise to a therapeutic level, at which point I could come home on Coumadin therapy for a minimum of six months. Additional blood work was ordered to check for genetic and/or acquired conditions that could influence clotting time, and they revealed that I do have lupus antibodies (lucky me!). Although I was not happy with this news, it will allow my health care providers to prescribe Lovenox before each future surgery, and decrease my future risk of another clot.

This hardly ever happens, but occasionally it does. Unexplained pain, redness, and swelling are warning signs that need to be taken seriously, even if we think "it's too late for that to happen. **Sally, 51 USA**

CHEST INFECTIONS

Chest infections are easily treated with antibiotics, but it's best if you don't develop them at all. Deep breathing exercises, and splinting your tummy with a towel or pillow while you cough, are good ways of reducing your risk of this complication (Figure 4.7). After surgery, if you can actively sit and take five really deep breaths in and out, at least three times a day, you can reduce the risk of developing a chest infection.

You are more at risk of chest infections after surgery if you smoke.

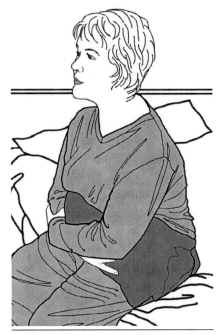

Figure 4.7: Splinting the tummy with a pillow when coughing.

NERVE DAMAGE

There are quite a few major nerves that run around the hip joint (see Figure 2.6 on pg 20) — the femoral, obturator, and sciatic nerves are the main ones — and they can be accidentally damaged during hip surgery. Problems like numbness and foot drop can occur. Foot drop is a significant weakness of the ankle and toes, making it difficult to lift your foot properly or to walk normally. If your leg is lengthened during surgery, these nerves will be stretched, which can cause various niggling problems, and take up to a year to adjust to. Even seemingly small changes in length (1 cm or less) can have a surprisingly large effect.

I got foot drop after my PAO. I saw a podiatrist so I could get the correct support for my foot, and saw a neurologist to find out the extent of the nerve damage. It is getting better. So far it has been just over 12 months and there is a significant improvement, but it has not resolved yet. Apparently it will take 18 months from surgery date to completely resolve, along with plenty of physio exercises and podiatry help. **Lea-Anne, 40 Australia**

URINARY TRACT INFECTIONS

Urinary tract infections (UTIs) are more common if you have had a catheter. However, they are easily treated with antibiotics and a good fluid intake. Cranberry juice has been proven *not* to have a significant benefit in avoiding or treating UTIs[49]. The first symptom of a UTI is often a painful stinging or burning sensation when urinating. The best treatment if you do get a UTI is to drink a lot of water, and take antibiotics. Urinary alkalinisers, such as Ural®, help turn the urine from acidic to alkali, which gives symptomatic relief. They reduce the stinging when urinating — but you will still need antibiotics as well.

WOUND INFECTIONS

Wound infections can be either superficial or deep. Despite all the precautions that are taken during orthopaedic surgery (such as extensive prepping of the skin with special washes, surgeons being scrubbed, laminar flow ventilation systems in operating theatres that reduce the number of organisms in the air, and being given antibiotics during surgery and in the early post-operative period), occasionally infections do happen. Most of the time it's just rotten luck!

Smokers are at increased risk of infection, compared to non-smokers. If you are diabetic, poor blood sugar control can also increase your risk of infection, so it is really important to get this reviewed and establish better blood sugar control before any surgery

Superficial Infection
This is where the infection is limited to the wound and surrounding soft tissues. It is recognised by increased pain and redness around the incision. Sometimes there is opening of the wound and leakage of pus, and you may have a slight temperature. This can sometimes just be treated with antibiotics, but occasionally needs to be 'washed out.' This is usually done in theatre under anaesthetic. They thoroughly clean all the infection away, and sometimes a drain is left in place, or a special type of vacuum dressing is used.

Deep Infection
Deep infection is a more serious problem, where the infection is in the joint itself. If any implants (screws, prosthesis, implant) are present, the infection can attach itself to them. Signs of this include: feeling generally unwell, increased pain in the joint that gets worse or doesn't resolve, a high temperature/fever, and all the signs of superficial infection. Unfortunately, a deep infection is really not good news, it can require prolonged treatment (weeks to months), with a combination of **intravenous** and oral antibiotics. Sometimes even removal and replacement of all the metal work or prosthesis is necessary. We discuss this sort of infection, and its treatment, in greater detail in Chapter 15.

This chapter has been pretty heavy going, we know. The rest of the book isn't so technical or confronting, so you've made it through the hardest bits! Please keep in mind that the incidence of complications is really very low, so even though the lists in this chapter may sound rather alarming, the chances that you will develop any of these is small.

Chapter 5
Disability and Pain

Living with an increased level of disability and pain is something you may well face with hip dysplasia. In this chapter we cover the experiences common to many hippies, and ways of dealing with the deterioration of your hip joints.

Whether you're on a long waiting list for surgery (as in the Australian public system, or the NHS in the UK), have a date for surgery, are holding off surgery for as long as possible, or aren't sure when to 'take the plunge' and have surgery, this time is not an easy one.

There are two likely scenarios here. Either you were diagnosed with hip dysplasia as a baby or young child, or diagnosed as a teenager or adult.

In both of these situations, your level of disability can increase gradually over the years, or you can see a sudden deterioration in your condition. The main difference is in what you're used to, and what you expect from your body.

If you've had a history of childhood hip problems, surgeries, braces, plaster casts, and all the other joys, you won't be all *that* surprised when your hip starts to act up again. Your childhood surgeon or specialist may have given your parents a bit of an idea of what you can expect in the future — this can include warnings about difficulties with pregnancy, and the expected life for any reconstructive surgery you've had done. However, this is not always the case, and you may find that even if your surgeries as a child were a 'success,' you can still develop problems that you weren't expecting.

If your hip problems have been gradually getting worse over the past few years, but you haven't had a diagnosis until now, it can be a relief to finally know what's going on. At least you can start to take some action at last.

In either case, you are at a high risk for developing osteoarthritis in the joint at a young age (under 55)[50]. Because your hip joint is abnormally shaped, it wears unevenly and osteoarthritis is the common end result[19].

Osteoarthritis

There are over 100 types of arthritis (which simply means 'joint inflammation'). Osteoarthritis (OA) is the most common variety, the 'wear and tear' sort, which is often seen in older age in the wider population. The cartilage covering the ends of bones becomes thin and breaks down. The ends of the bones are left uncovered and unprotected, so it becomes harder for the joint to move smoothly (see Figure 5.1).

Arthritis that has been caused by another condition (in this case, hip dysplasia) is known as *secondary osteoarthritis*. The increase in pain in your hip over time is generally due to secondary osteoarthritis (OA), although it is possible to have pain from a dysplastic joint without any signs of OA. In a dysplastic hip joint, the uneven and abnormal shape of the acetabulum and/or femur head can speed wearing down of the cartilage.

When the ends of the bone are exposed, they get inflamed, and can be stimulated to grow small bony protrusions, called bone spurs or osteophytes. These can be seen on an x-ray. They limit your range of

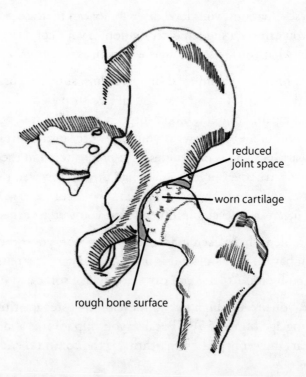

Figure 5.1: Osteoarthritis of the hip.

motion, and cause pain by rubbing against soft tissues and bones. Due to wear and tear, loose bodies can appear within your hip joint. These loose bodies are tiny bits of cartilage or bone, or other tissue, and can float about within the joint space. They can occasionally get caught in between the acetabulum and femur head, causing the joint to 'lock up' temporarily. They are a part of OA. The joint space also narrows. Osteoarthritis is generally diagnosed by having an x-ray and a physical examination.

The main symptoms of osteoarthritis are aching pain and stiffness in the joint, which is often worse when moving after sitting, but which improves with rest. Pain can be worse later in the day, and generally worsens the more the joint is used. Sometimes joints can feel warm or a bit swollen. Your range of motion may be decreased. When OA is advanced, the joint can hurt most of the time, even when you're resting. With hip osteoarthritis, you will find that twisting movements, and weight-bearing (standing, walking, and so on), are the most painful.

In general, hip arthritis develops very slowly over many years, but it can also suddenly deteriorate. It's quite common for things to be bearable on the whole, but then suddenly get much worse. This may be due to a piece of cartilage or the labrum tearing acutely, or a flare up of inflammation (this might settle with anti-inflammatories), or because the joint is no longer able to compensate for your deformed anatomy as well.

There are many books and good web sites out there for more detailed information on osteoarthritis. Looking up your national arthritis charity is a good place to start (Arthritis Australia, Arthritis Care (UK) and Arthritis USA are just a few). However, you may find that they do not cover hip dysplasia, or that their target audience is a somewhat older group.

You might like to visit these web sites:

◊ **Arthritis Australia** < www.arthritisaustralia.com.au >

◊ **Arthritis Care** < www.arthritiscare.org.uk >

◊ **Arthritis USA** < www.arthritisusa.net >

Increase in pain

When your hip is getting worse, there is often collateral damage. You favour your painful hip, start limping, and put more stress on your 'good' hip. It's a natural and unconscious reaction.

Most of the time, the resulting problems will be temporary, and will ease up once you've had surgical treatment for your DDH. Sometimes the threat of further deterioration of *other* joints (such as your 'good' hip, or your knees) can help push you towards a decision to have surgery.

The following are some of the main complications that can develop as a result:

BURSITIS

Bursitis is the inflammation of a **bursa**. What's a bursa, you say? Good question! A bursa is a small sac of jelly-like fluid that cushions an area of our body where there is friction and movement. So bursae are common around joints, sitting between bones, and the muscles and tendons that glide over them. Some of the risk factors for developing hip bursitis are having previous hip surgery, rheumatoid arthritis, overuse, and leg length differences.

With hip problems, the main bursae that are affected are the greater trochanteric bursa and the iliopsoas bursa. Figure 5.2 illustrates the major bursae around the hip joint. The greater trochanter is the 'point' of the hip bone that you can feel on the outside of your thigh, so trochanteric bursitis will hurt on the outer side of your hip. This region is the attachment point for a lot of muscles that move your hip, so if the bursa becomes inflamed, any movement of your hip muscles will further irritate the bursa and cause pain. Leg length differences can make you more likely to develop this sort of bursitis.

I developed pain on the outside of the hip after a long weekend in Paris with lots of walking, but no specific incident. I thought I'd pulled a muscle or similar, but it didn't get any better over a week or so. Very quickly it became very sore, to the extent I couldn't stand for any length of time, and walking became difficult, and it felt bruised and sore to the touch. I believe trochanteric bursitis is often known as a runner's injury/condition, but in my case I think the weird anatomy that goes with hip dysplasia was the cause. **Deirdre, 36 UK**

Iliopsoas bursitis is less common, but can be caused by hip joint abnormalities such as DDH. The iliopsoas bursa is positioned on the anterior (front) side of the pelvis, under the iliopsoas muscle, in front of the hip joint, deep in the groin. The iliopsoas muscle flexes the hip — the sort of movement you make when kicking a ball. This type of bursitis is felt as sharp pain deep in the groin, the front of your hip, and possibly running into your upper thigh. Your hip may feel stiff and weak, and like it's clicking, collapsing, or snapping. The pain may be intermittent, and can come on suddenly with certain movements.

It's definitely worth seeing your doctor or surgeon if you're experiencing any sharp pain in your hip, especially if it's a new symptom. Bursitis can be treated with rest, avoiding activities that irritate the area, use of a walking stick or crutches, NSAIDs, heat, ice, and stretches (see your physiotherapist).

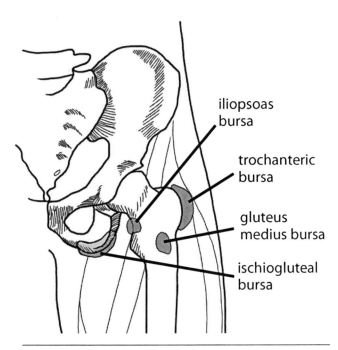

iliopsoas bursa

trochanteric bursa

gluteus medius bursa

ischiogluteal bursa

Figure 5.2: The bursae around the hip joint.

If the problem persists, cortisone is an effective treatment, injected directly into the bursa. The doctor often uses ultrasound to see exactly where the needle is going, so the injection goes to precisely the right spot. As you may have guessed, this is not an enjoyable treatment. A local anaesthetic is given, but for deep bursitis (especially iliopsoas) it is still pretty horrible. Despite this, a cortisone injection can be *very* effective, and should settle everything down rapidly. See Chapter 3 for more details on steroids.

There are no hard and fast rules about how many cortisone injections you can have; several in a year probably won't present any problems. If one or two injections into the area aren't helping, it's unlikely that further injections will help. Repeated and frequent injections of corticosteroids into a region of the body can damage the tissues over time.

Another delivery option for steroids is **iontophoresis**. In this transdermal technique, cortisone is driven through the skin using electricity. Two electrodes (one anode and one cathode) are placed on your leg or hip, one of which contains cortisone in a pad. A low electrical field is then set up, and the cortisone moves painlessly through your skin into the affected area. The only sensation you feel is slight buzzing or tingling in the region of the electrodes. Obviously, this isn't a quick a fix like an injection — you may need as many as five or six sessions, each several days apart, and each one taking roughly 30 minutes. But it's half an hour lying down reading or resting, while electricity does its thing. It may not be an option for more severe cases of bursitis, but it's definitely worth asking about.

I have had several bouts of iliopsoas bursitis in my right hip; the first one was before my left THR, and was treated very successfully with an injection of cortisone into the bursa (which was an unpleasant experience, the needle has to go so far in!). My second bout was after my THR, affecting my right hip again. It was initially treated with iontophoresis. Around this time I was put on high doses of oral corticosteroids for an unrelated problem, so I think that helped the bursitis as well. **Denise, 47 Australia**

If your bursitis is resistant to all these treatments, and causing you significant disability, there are options including surgery available, but you need to discuss your case individually with your surgeon or physician.

TENDONITIS

Tendons are tough flexible bands that attach your muscles to your bones. There are many tendons around your hip joint, of all sizes and shapes. They can become inflamed, a condition called **tendonitis**. The normal smooth action of the tendon doesn't work, and movement becomes painful. Tendonitis is generally an overuse injury, but can also be caused by physical deformities of the skeletal structure — such as in dysplasia. The area will be tender, the tendon itself may be swollen, and it will hurt to move muscles in that region.

Treatment for tendonitis is similar to that for bursitis — rest, ice, NSAIDs, and cortisone injections if necessary. For stubborn cases, surgery may be required.

LABRAL TEAR

The labrum is the cartilage that forms a rim around the hip joint. It makes the joint deeper and helps to stabilise it. When a dysplastic hip starts to deteriorate, this rim of cartilage is at risk and can tear. As with bursitis, the symptoms are commonly groin pain, limited range of motion, and clicking or snapping sensations in the hip joint. While rest, NSAIDs, physiotherapy, and cortisone may help with relief of symptoms, arthroscopy is the main treatment option for labral tears. There is an in-depth discussion of this procedure in Chapter 4. See Chapters 1 and 2 for more details of the anatomy of the labrum.

KNEE PROBLEMS

Your knees can take quite a battering when your hips aren't functioning well. Hip pain can also be referred to the knees, so some of your knee pain may actually be coming from your hips! This is because the nerves that supply the hip joint also supply the knee joint.

It's not uncommon for knees to become swollen and sore, especially when you're limping badly, or on crutches. Be gentle on your knees, and especially avoid kneeling on the ground or other hard surfaces.

See your GP if you're having any problems with your knees. Your physio may be able to help with strengthening exercises and supportive strapping. A baking soda poultice can help with swelling (see Chapter 19).

Mobility and functioning

All of the above conspire to reduce your mobility and general functioning. It hurts to walk, let alone run. Sitting can be painful, sleep becomes a nightly battle, and your sex life a thing of distant memory! You may have to temporarily stop many of your favourite activities, from sports and other physical activities like dancing and gardening, to more mundane things like shopping at the mall, where a lot of walking is required, and playing with your kids at the park, or even bathing them.

As pain and disability take a greater toll on your family, work, and social life, it's no surprise if you succumb to depression and frustration. It's important to focus on what you *can* do, and find things you can still manage and enjoy. We discuss this in more depth in Chapter 7.

SOME BETTER DAYS

Joints being what they are, there will be ups and downs, times when all these bothersome symptoms abate, even for weeks at a time. While it's great to have some respite from the pain, these good times can really make you doubt whether you really need surgery, especially once you've got a firm date for surgery! It's quite normal to have these doubts, and to feel relieved when all the pain returns – as it almost definitely will, we're sorry to say.

If my pain ever goes away for a split second, I think to myself, "Am I overreacting? Is my pain really as bad as the doctors describe it?". **Bri, 15 USA**

What you can do to help your condition

The following things may help to improve your level of functioning, and pain control, and are certainly worth trying.

WEIGHT LOSS

If you're over your healthiest or most comfortable weight, your doctor or surgeon may advise you to lose weight. *Sigh.* The reason for this is that carrying less weight makes a dramatic difference to the stress put on your joints. For every kilo or pound, your knees are under three to five times that many kilos or pounds of stress, and up to six times the pressure on your hip joints when you're walking. So if you lose 10 kilos, that's 60 kilos less stress on your hip joints every time you take a step. (Losing weight also reduces stress on hip replacements, and can help extend the life of your prosthesis.)

This is much easier said than done, as we all know! And the whole process is only more difficult when you're also dealing with serious hip pain, and related problems. The last thing you're able to do is go for a jog, let alone a brisk walk.

Dietary change is important for weight loss. The absolute basics are to eat less and exercise more. Less energy in, more energy expended. That's it.

Some Crucial Elements Are:

◊ Eat a balanced diet, not cutting out any of the major food groups.

◊ Look at reducing your portion sizes. Smaller plates can help.

◊ Look for ways to reduce your saturated fat and sugar intake.

◊ Make changes gradually, rather than going 'cold turkey'.

◊ Avoid fad diets. They don't work. That includes restricted foods, blood types, and thousands more!

◊ Cut back on take-away foods, cook your own meals.

◊ Increase the amount of fresh fruits and vegetables you eat.

◊ Steam, roast, poach, bake, grill, and microwave instead of frying.

◊ Think of rich treats like chocolate, chips, and cream as 'sometimes' food; don't make them 'bad', but restrict how much you have, and make them occasional treats, savouring every bite!

◊ Avoid buying lots of 'sometimes' foods — if they're not in the house, you won't snack on them!

◊ Become aware of and cut back on non-hungry eating.

◊ Have one carb-free meal a day, if you can, and try to have a bit more protein (tin of tuna, egg, low-fat cheese, yoghurt and other dairy, popcorn, oatmeal, beans, meat, oatmeal, poultry ...).

◊ Search online for recipes, there are a plethora of web sites for "low-carb" and "high-protein" meals.

◊ A diabetic diet is often a good one to try (low GI, low carb).

◊ Drink water, sugar-free drinks, and (limited) fruit juice instead of soft drinks/soda.

◊ Add weights training into your exercise program. They're nice and low-impact for your joints, and are very effective! Avoid moves which hurt your hips or knees.

◊ Keep a food diary, Many studies have shown that using a food diary helps people to lose twice as much weight, and keep it off for longer[51].

We highly recommend the book *If Not Dieting, Then What?* by Dr Rick Kausman < ifnotdieting.com.au >, and web sites Weight Loss for Busy People < www.weight-loss-for-busy-people.com > and Livestrong.com <www.livestrong.com>. The free iPhone/iPad app *MyFitnessPal* < myfitnesspal.com > is excellent too, if you want to keep a food diary.

PHYSIOTHERAPY / PHYSICAL THERAPY

A good physiotherapist, or physical trainer, is worth their weight in gold. They can help you with exercises and stretches to help manage your condition, and provide some relief for your symptoms, even though they can't correct the underlying cause (that's what your surgeon is for). Ask around your friends for a personal recommendation, or ask your doctor, if you don't already have a physio, or are unhappy with your current one.

ALTERNATIVE THERAPIES

Chiropractic, osteopathy, acupuncture, and massage therapy are just some of the wide range of alternative therapies that are available. Whilst medical evidence and opinion is divided about the effectiveness of these, we appreciate that some people get a lot of benefit from them. What we will say is if you do feel something 'alternative' is helping, make sure the treatment is as an *adjunct* to standard medical treatments and specialists, not instead of it. Hip dysplasia is a structural problem. Chiropractic and osteopathic treatment *cannot* correct it. Always check with your GP, surgeon, or physiotherapist before starting anything new.

EXERCISE

Exercises can help with your general fitness level, and will help with losing weight too. It's important to keep your joints moving, despite the pain. Check with your doctor and/or physiotherapist before embarking on a new exercise routine, to check if they have any advice or guidelines for your particular situation. Low-impact exercise will probably be the best for you, so look at things like cycling, an exercise bike, walking, seated exercise programs, and swimming.

Swimming and Hydrotherapy

Your physiotherapist should be able to advise you on good exercises to do in the pool, as well as swimming. Walking backwards and forwards in chest-deep water is a great start. You don't even need to be able to swim.

The book *Heal Your Hips: How to Avoid Hip Surgery – and What to Do If You Need It* by Dr Robert Klapper and Lynda Huey has an excellent hydrotherapy program you can do by yourself in the swimming pool, aimed at people with hip problems. Their web site has a good amount of information too < www.hiphelp.com >.

You can join a hydrotherapy class, these may be run by your local arthritis support group or hospital. They have the advantage of being at a set regular time, and run by a trained instructor. Be prepared to be one of the youngest in the class!

Walking
If you're able to walk small distances, even if you're using a walking stick or crutches, doing this every day or two will help your fitness level.

Exercise Bikes and Elliptical Trainers
You may have one of these gathering dust in the garage, or pushed into a corner — bring it back into active duty, even if you just use it for five minutes a day on the lowest setting. If you don't want to buy a new machine, look for a second-hand one, borrow one, rent one, or sign up to a gym.

Pilates and Core Stability
"Core stability" refers to the strength of your body's "core" — which is comprised of your abdominal wall, pelvis, lower back, and diaphragm; together they support your spine and pelvis. Us hippies often have poor core stability.

Your physiotherapist or Pilates trainer can help you learn how to improve your core stability, and strengthen your abdominal muscles. If you decide to get into Pilates, it's important that you seek private tuition, at least initially. Ask for a strengthening program, and explain your hip problems. The last thing you need is to be in a big group, and falling behind the pace, with exercises that are beyond your current limits, and don't take into account the state of your hips and knees.

Other 'Arthritis-Friendly' Exercise
Other low-impact exercises include gentle yoga, weight lifting, tai chi, and rowing machines.

Pain relief

The first thing to do with pain relief is to see your GP about it, and be honest about the impact pain is having on you. You may need a medical certificate for work or school, explaining that you need to use the lift, or that you need to be excused from sporting events at school, or that you need some time off work. Your GP can also provide pain relief medication on prescription, if over-the-counter medications aren't cutting it for you any more. NSAIDs such as Celebrex can be very effective, and many are safe for long-term use.

See Chapter 3 for more details on the pain relief ladder, medication, and keeping a pain diary.

HEAT AND ICE

Heat can provide a lot of pain relief, used in moderation. Hot baths can be wonderfully relaxing and ease your aching joints. Don't forget the bath salts and candles! Adding Epsom salts to the water will help reduce 'prune skin', and can ease inflammation.

There are a range of heat packs available, from wheat or rice packs which are heated in the microwave, and the good old standard hot water bottle, to chemically activated heat packs (containing sodium acetate). There are instructions on how to make your own heat packs in Chapter 19.

Ice packs are another good standby, reducing inflammation and pain. You can just bundle some crushed ice into a plastic bag, wrapped in a wet towel. There are a wide variety of gel packs which you simply store in the freezer, and you can make your own (see Chapter 19). Never put an ice pack directly onto your skin, wrap the pack in a cloth. See Chapter 3 for more details on using ice packs.

With both heat and ice packs, the main thing to remember is not to leave them on your skin too long, not longer than 20 minutes at a time, and no more than three times a day. Alternating heat and ice packs may work well for you, too — it's worth a try, anyway!

TIPS AND TRICKS FOR PAIN RELIEF

You may find that these techniques and things may provide some help with pain relief for you.

◊ Analgesic creams (likely to be of limited help).

◊ Tiger Balm and other 'heat producing' creams.

◊ Mindfulness and meditation, more on this in Chapter 7.

◊ Sleep and rest — pain will feel worse if you are overtired.

◊ The book *Manage Your Pain* by Nicholas, Molloy, Tonkin, and Beeston (from the Pain Management and Research Centre at the Royal North Shore Hospital, Sydney, Australia) has a thorough treatment of this subject and practical advice (as do many other books and web sites). < www.painmanagement.org.au >

I tried to manage pain relief pre-surgery with medication and managing what I did. For example, if I did nothing and rested, sometimes that would be more painful than moving around, so I tried to get the best balance for me of sitting, lying, moving, swimming, and doing mild exercise. I used the jacuzzi, which helped the pain. I also used a hot water bottle and Tiger Balm on the hip which seemed to help too. Medication-wise, I was taking paracetamol regularly, diclofenac for anti-inflammatory, and lansaprozole as it upset my stomach, and tramadol when I needed it. Some days I went with no tramadol, and other days I needed up to four. **Sian, 29 UK**

We hope that this chapter has given you some positive coping strategies for the ongoing and worsening physical problems you may be having with your hips. Hopefully you have gained some things to think about when to schedule you surgery, if that is a path you're taking. In the next chapter we start to delve into the practicalities of living with increasing disability.

Chapter 6
The Practicalities of Disability

Apart from the physical problems associated with increasing hip problems and disability, there are a bunch of practicalities that you need to deal with. The ones that we discuss in this chapter may help make your life easier (although we're not sure if health insurance falls into that category!).

There are many disability aids that can help you as your hips get worse, and during the recovery period after surgery. Walking sticks, crutches, grabbers, **sock gutters**, and a variety of long-handled tools can make your life easier.

If you haven't developed osteoarthritis in your hip, you may be a good candidate for a PAO (see Chapter 4). In this case, you may be in the strange position of not being in much pain at all, but signing up for major hip surgery. You probably won't have any need for these disability aids before surgery — just for some months after it. We discuss disability aids in detail in Chapter 11.

Disability parking

A disabled parking permit, or handicap placard, can make a huge difference, and when you have hip problems, you may qualify. It is one of the very few benefits of your increasing disability.

Getting around will become more and more difficult, especially as osteoarthritis takes its toll. Some people need a permit for years before they have surgery. You also won't be allowed to drive for some weeks or months after a THR or PAO, but it can still be helpful to get a permit so whoever is driving you around can park close to your destination. In some places, a disabled permit also allows you to park in 'regular' car parks for free, which can be a huge help — the last thing you want to have to worry about is limping back to your car by a time limit to avoid a fine!

The procedure for obtaining a disabled parking permit differs between countries, and even within a country. It's best to contact your local department of road services, and ask what you need to do. Be aware that disabled parking permits are notoriously difficult to get in the United Kingdom.

In general, there will be a form that you take to your GP or surgeon, which they fill out, specifying what your disability is, and how long you will need the permit for. The transport department will then (hopefully) issue a permit, which may be free (as in some Australian states), or have a small fee attached to it.

Permits can be issued for as little as a few weeks to cover post-op times, to as long as "indefinite". The best advice we can give you about this is to start the application process *months* before your surgery, as bureaucracy is rarely fast, and you have to get your doctor or surgeon to fill in forms.

Unfortunately I was slow and filed for it a bit too late (about two weeks before my op), so by the time they did it I was already about three weeks post-op. I would suggest filing for it as soon as you get the necessary documents. I have appreciated and enjoyed being able to use the disabled parking, since I have been on crutches for over a year. It has helped me in times of rain, snow, and just in need of a good parking space!! **Arpine, 28 USA**

I asked before surgery and my surgeon said no. But after surgery, I took the papers to his office and asked if he would sign them, and he was happy to. I don't know if this was because of the complications I had with foot drop, or if he just had a change of mind? **Lea-Anne, 40 Australia**

I applied for one, but was rejected. **Deirdre, 36 UK**

I downloaded a form from my local council web pages, completed and sent it off with a cheque for £3. They sent a form to my GP to complete, then sent me a badge which is valid for four years. The whole process took about five weeks. **Sian, 29 UK**

I applied for a disabled parking permit two weeks after my surgery. My OS would have approved one before my surgery, but he forgot, because he's primarily a pediatric surgeon, so it's not something he deals with often. **Melanie, 29 USA**

At first I was upset that I had a permanent permit. Then I thought ... I have RockStar parking, always! Not so bad after all. I don't abuse it and use it with care. I'm still using a cane after having my THR surgery over a year ago, so the close parking does come in handy. When I first received my permit, I could never find a parking space ... lots of disabled people out there. **Ina, 41, USA**

Make sure that you lock your car securely, too – it's not unknown for disabled parking permits to be stolen. The last thing you want to have to deal with soon before or after surgery is sorting out a replacement permit (and a fine for having apparently parked illegally!).

Someone stole my handicap placard from my car while I was visiting my mother-in-law! How disappointing is that?! Shame on me for not locking my car door. **Marcie, 30 USA**

Coping without a permit

If you're unable to get a permit, here are some tips on how to manage:

◊ The main thing is to ensure you've got room next to the car, so you can open your door wide, especially if you're on crutches.

◊ Park at the edge of a parking area, next to an open space or the curb, so when you open your door, another car won't hem you in.

◊ If you're a passenger, ask the driver to stop partially out of the parking space first, so you can get in or out when there's more room to open the door wide, and manoeuvre yourself into or out of the car.

◊ Park as close to your destination as possible, of course.

◊ If you're a passenger, ask the driver to drop you off at your destination, and meet up with them once they've parked the car and done all the walking back to you.

◊ Consider using public transport, if you're steady on your feet, or use taxis, if you can afford them.

I used the disabled parking spots, particularly at the supermarket, when I was on two crutches, just because it made it easier to get in and out of the car. I didn't have a disabled parking badge, but I was never challenged about it. I did try a normal parking spot, and found I couldn't get back into my car because some idiot had parked too close for me to open the door wide enough to get back in! LOL! Not that they were to know I was on crutches! **Annick, 47 UK**

What are you parking there for?!

Unfortunately, some people can be downright rude about disabled parking. There are two main problems (which are directly related!): people without permits using the disabled spots, and people challenging *your* right to use them.

There's not a lot you can do about inconsiderate able-bodied people using the parking places. Some people like to carry a few little printed signs, saying things like 'Stupidity isn't a disability', to put under the windscreen wipers of offenders. There's even a web site:
< www.apparelyzed.com/parking >

Let's just say it was probably a good thing I wasn't to quick on my feet with regards to the many very able-bodied people who just popped into a disabled bay for convenience, stopping the real people who have a need to use it! I did occasionally lean across to the wheel, and beep the horn flashing my disabled badge through the window at them like I was a police officer or something. They always slunk sheepishly into the car with slightly rosier cheeks!
Rhianna, 28 UK

If you see an apparently perfectly healthy driver returning to their car, make sure you've checked their car (windscreen, rear vision mirror, rear licence plate) for a disabled parking permit. If they really don't have one, then you may feel like saying something. Remember, though, that many disabilities are invisible, and some able-bodied drivers may have a disabled child or elderly parent they drive around.

The bigger problem for us hippies, in general, is that people accost us demanding to know what right *we* have to park in these spots — this seems to be more of a problem for younger drivers.

I had an upsetting experience with using a disabled bay. We parked outside a children's play centre and the space was immediately outside the main door – excellent! I displayed my badge in the windscreen as usual. When we returned to the car there was a note on the windscreen saying "Would you **please** adhere to the parking restrictions".

I went straight back in to the receptionist waving my blue badge, and said would they please check next time they put a note like that on a car as I found it very upsetting. It actually made me feel really bad that they thought I had parked in a disabled bay without a blue badge. If I can park my car close to the entrance of any place I'm visiting without using a disabled bay, I do so in order to leave the disabled bay free for other people who need it.

I've never been challenged by a member of the public about parking in a disabled bay. I have thought about dropping my trousers to show the 12" scars running down both legs and groin, but that's probably a bit drastic!
Claire, 30 USA

First of all, ensure you have your permit/badge clearly displayed. Secondly, keep your cool. We know it's tempting to defiantly reveal your impressive scars to the world, but it's really not necessary, or appropriate when many hip scars are in the groin, we don't need you getting arrested for indecent exposure! Just affirm that you are allowed to park there, and that you have hip problems (if you feel like going into even that much detail).

Of course after I got off crutches I was getting looks— sometimes I had to fake the limp to avoid them. But I knew I deserved that placard, and didn't want to explain to every person looking at me what I've been through, and why I have it. A few times colleagues at my work commented on my parking on one of handicap spaces, "Why are you parking there, you're not handicapped any more?" That was unpleasant. But overall it was good to have it. It definitely helped — especially because I had my surgery in the late fall, so was on crutches for the most of the icy-slippery winter, and it was great to not have to crutch through the icy parking lot to my destination. **Arpine, 28 USA**

It will be obvious that you're entitled to be there if you're on crutches, but it can be a bit more tricky if you can *mostly* walk okay, but are suffering from pain and fatigue — difficult things for a passerby to notice. A good strategy is to keep a walking stick in the car, even if you don't desperately need it. Grabbing your stick when you get out of the car will keep the nasties at bay.

If I think people are looking at me in a judging way I just limp more!
Lea-Anne, 40 Australia

Only use the disabled parking places if you really need them, too. Once you're walking and moving around more easily, see if you can manage with parking in a regular spot (even if you're entitled to use the disabled spots), so those in greater need can use the disabled places. The extra walking is good exercise too!

I think even though we may have a permit, it doesn't mean we should use the parking if we don't really need it (if there are plenty of other car parks close by), I like to leave it to the elderly or those who really need it. **Lea-Anne, 40 Australia**

96

Driving

If driving is causing you pain, or your legs are weak, and you think this will affect your ability to brake or change gear, then you really need to consider whether you should be driving at all. Some strong pain medications can affect your reaction times and ability to think clearly, too.

From a legal perspective, there is no rule or law (that we know of, anyway!) that says if you have hip dysplasia you cannot drive; this is also true for arthritis. Do check with your local road transport authority, though, as rules do differ across states and countries. However, you may find that you will not be covered by your insurance if you have an accident, and it can be shown that your hip problems affected your driving. If you can drive comfortably, then there's no problem. It is best to discuss things with your doctor and insurance company, this is true for post-surgery driving too.

The DVLA (the licensing authority in the UK) says that if you have severe arthritis, particularly with pain, adaptations to your vehicle may need to be made. If you do make adaptations to your car, you need to inform them.

Out and about

Just because you're increasingly disabled by your hip problems, it doesn't mean you have to stop your social life! It will just take a bit of adjusting, some compromises, and more planning to achieve. These tips and tricks will help you out:

◊ Plan ahead — know how much walking/exercise will be involved.

◊ Tell your friends, so they know what you're able to cope with, and what's too much.

◊ Check access to places if you have mobility problems.

◊ Avoid venues with lots of stairs, check for elevators, ramps, and seating.

◊ If you are worried about pain, take a walking stick or crutches with you; it's better to have them and not need them, than be stuck without when you do need them!

◊ If pain medication helps you, take some just before you go out, and have an extra dose with you just in case.

◊ Plan the journey home — bearing in mind you may well be more tired than when you set out.

◊ Set time limits, to avoid overdoing it.

◊ Try not to get completely smashed, you need your wits about you to avoid falling over, and hurting your hip even more! Crutches and alcohol don't mix well. And they're not a good mix with pain medications either. Keep alcohol consumption to a minimum.

When I had to give in and use crutches I would go out with my girls for the night drinking and clubbing and the first thing I would look for instead of the bar was a table with seating, and whilst I was sitting down in the club, my ever-so-loyal friends would make a semi-circle round my table and dance there to me all night long, trying to make me feel part of the night. When all I remember thinking is I want to burn these crutches, put on a massive pair of stupidly high shoes, go straight to the bar, and get drunk and dance with my girls!

I started to feel sorry for myself that my first thoughts of going out for the night were "Will there be steps? Will there be seats? Will I be able to make it to the end of the night without letting on how much pain I'm in?" And what I realized from all of that was how lucky I was that my girlfriends had stuck by me from school, they always looked after me, included me in everything they were planning, and even started to check ahead that certain places and venues would be okay for me to go into! These amazing ladies and my boyfriend were and still are always there for me! You really do very quickly learn who your true friends are!!
Rhianna, 28 UK

CONCERTS AND SPORTING EVENTS

You will find that most of the big venues offer special arrangements for people with mobility problems. Some of them require you to be registered as 'disabled', however for those with short term mobility issues (eg after surgery) if you provide a doctor's letter they will usually accept this. These special tickets are usually good seats, and your carer is often given a complimentary seat, so in effect you are getting two for the price of one.

After my second PAO I was using my wheelchair to get around. Bon Jovi was playing at a local venue, and I was desperate to see them, but knew that I couldn't manage sitting in the small seats or standing for the concert. We contacted the venue and they offered us disabled area seating, where I could sit in my wheelchair, and a free ticket for my carer (my husband). The view was amazing, we had such a great time, and it was all so easy to arrange. We went again three nights later! **Sophie, 29 UK**

Health insurance

NHS VS. PRIVATE INSURANCE IN THE UK

The National Health Service (NHS) is one of the most complicated and variable systems out there. The good news for Brits is that — although it may take time, it can be frustrating as hell, and it is not a system without its flaws — ultimately you *can* have your surgery, and you don't have to pay for it.

There is a lot to be said for the NHS, especially if you have other medical problems. There are numerous other specialists available in case things go wrong. The social support systems are very good, and physiotherapists and occupational therapists are usually very experienced in dealing with patients with DDH (which is not always true in the private sector).

If you are lucky enough to have private medical insurance, and are fit and healthy, you may decide to have your operation privately. This means nicer hospital stays, quicker outpatient appointments and investigations, and potentially frees up NHS space for those who cannot afford it. It is ultimately your choice if you are in that position.

INSURANCE IN THE USA

What a minefield. Neither of us is based in the USA, so we have limited experience to draw on. However, we have gathered a few things from our research and from American hippies:

Coverage varies drastically depending on your insurance company and the scheme you are on within it.

◊ You may need to fight for your case, especially as hip dysplasia is only managed in a few specialist centres. Many American hippies end up having to argue their case for the need for surgery with their insurer, and fight to get the best surgeon for the job.

◊ There is often a large discrepancy between the surgeon and hospital bills, and what patients actually receive as a rebate.

◊ If your hip dysplasia is deemed to be a 'pre-existing condition' you may have more hassles.

◊ It can be incredibly stressful!

Private health insurance is the main coverage in the States, and can be very expensive and complicated. Health insurance is often offered through your employer. Members pay a monthly premium, which gives them access to a network on healthcare providers and doctors.

There are four main types of insurance options : HMO (Health Maintenance Organisation), POS (Point of Service plan) PPO (Preferred Provider Organisation), and Indemnity. If you are in an HMO, you can only be covered by seeing a provider within that network. In an Indemnity plan there is no network, so you can go to any provider you wish, and services are covered to an agreed percentage (e.g. 70% of costs). Within POS and PPO plans, it is much more cost-effective for members to see doctors within the network rather than out of it.

The other issue is there are many jobs in the States without insurance benefits. That means you have to apply for individual health coverage, unless your spouse has family coverage through their work, or you're 26 or younger and can be covered by a parent's insurance policy.

If you do apply as an individual, then the insurance companies love to deny people who have pre-existing health conditions. Basically, the insurance companies look for any excuse to deny coverage to all but the healthiest, least-expensive patients.

100

This won't change until 2014 when federal law will prevent it. In the meantime, individuals who can't get coverage due to pre-existing conditions have to go into a special high-risk, expensive insurance pool.

However, don't lose heart. If you have good insurance, everything tends to go smoothly, and there are no problems if your tests and so on are approved ahead of time. Often the staff at the doctor's office or hospital can help you with this, and will offer to do so.

The USA Medicare system allows for free coverage for low-income and retired people, but not all doctors will see Medicare patients because of the low rates of reimbursement.

With an HMO, you have to get a referral before seeing a specialist, but I've been fortunate enough to avoid those plans. My current plan is very similar to an HMO, except that I do not need to have a referral and I have two levels of network coverage (in addition to the possibility of going out of network).

Another issue is when you have two insurance plans, as I had for a while. I was covered under the military insurance as the spouse of a retired service member, so when I had arthroscopies, I paid almost zero because the military insurance picked up the balance. However, the military insurance *had* to be the secondary coverage, and I had to send several of the primary insurance Explanation of Benefits forms to get things covered. I believe some of the other USA government insurance programs (Medicare, and so on) work the same way if you have a commercially obtained policy. **Louise, 45 USA**

MEDICARE AND HEALTH INSURANCE IN AUSTRALIA

It is perfectly possible to have the surgeon of your choice, free surgery and a free hospital stay within the Australian Medicare system. Well, your taxes *are* paying for it ...

The main downside of using the Medicare system is the length of the waiting list, which can be as long as a year or more for non-urgent surgery like hip dysplasia. If your condition deteriorates, your surgeon can change your priority listing, and get you moved up the list. Most surgeons will only have a few spots for public patients each week, with much more availability for private patients.

If you have private insurance, you may actually have a fair few out of pocket costs — as well as your monthly premiums, you will be billed for the surgery and hospital stay, and only receive a portion of this back from your health insurer (depending on what coverage plan you have, of course). You may have a better chance of getting a private room, or a two bed room, instead of being stuck in a four bed ward in hospital, but there is no guarantee of this. The main advantage of going private is being able to schedule your surgery relatively quickly, and fit it in around your own schedule. With Medicare, you will have to drop everything to fit in with the date the hospital assigns you, and if you refuse a surgery date you're likely to end up at the bottom of the waiting list again.

Before deciding to use private health insurance, it's a good idea to ask the doctors (surgeon *and* anaesthetist) and the hospital what they will be charging, and then check with the insurance company how much they will pay. You may be able to check such things by logging on to your health insurance company's web site.

Private insurance means you can also have the choice of going to a public hospital or to a private hospital (so long as your preferred surgeon goes to that hospital), which may or may not be a better option than having surgery and recovery in a public hospital.

If you choose to have surgery as a private patient, apart from being able to have your surgery more quickly, and at a time of your choosing, you might be freeing up a Medicare space for someone else.

I went through the Medicare system for my hip replacement, as I can't afford private health insurance. I had the choice of the surgeon I wanted, and was on the waiting list for about a year. My nine day hospital stay, all surgery, and post-op physio within the hospital, was entirely free. I was in a four bed ward, which was admittedly a pain, but the price was right! The main problem I had was the hospital administration kept losing my surgeon's requests to increase my priority category, to get my surgery happening more quickly as my condition deteriorated — and that wasn't a Medicare issue, but a hospital bureaucracy problem. **Denise, 47 Australia**

OTHER COUNTRIES

The rest of the world — we're afraid that if we went through the different protocols in every country one by one, it would probably result in the world's longest and most boring book. Consult your surgeon and local health insurance companies to find out what sort of coverage is offered for orthopaedic surgery.

We are not trying to influence your decision as to how to proceed with regards to health insurance, that ultimately needs to be between you and your surgeon. However, we hope we have been able to give you the means to ask appropriate questions and provide some background information.

Government disability benefits

This is a brief summary of the general availability of disability benefits in the United States, United Kingdom, and Australia, at the time of writing (2011). While you and your carers may not qualify for any benefits, it doesn't hurt to find out what the situation is in your country. You may be pleasantly surprised!

Social workers can help you to work out what you might be eligible for, and how to access help, including financial assistance and various practical services.

BENEFITS IN THE UK

It is worth contacting someone at your local job centre, or go online, as you may be entitled to some benefits. This can include ½ price bus/tram fares, Employment and Support Allowance, Disability Living Allowance, and Carers Allowance; the list does go on. Eligibility is based on personal circumstances.

ACCESS TO WORK (UK)

This is an organisation that can help with transport to and from work, and financial advice with regards to returning to work.
< www.accesstowork.co.uk >

BENEFITS IN THE USA

Although not typically seen as a 'welfare state', for those in America there are several options that might be available to you if you think financial issues are going to be a problem.
< www.usa.gov/Citizen/Topics/Benefits.shtml >

BENEFITS IN AUSTRALIA

You may be eligible for some disability support programs through Centrelink, such as the Disability Support Pension, Mobility Allowance, Sickness Allowance, or the Youth Disability Supplement (if you're under 21). The Disability Support Pension provides an income if you are deemed to be unable to work for two years because of illness or disability. The Mobility Allowance can help if you're unable to drive, and can't use public transport to get to work or training. The Sickness Allowance provides support if you're temporarily unable to work because of illness, injury or disability, and includes cover for self-employed people. Most of these benefits are subject to income and assets tests (apart from the Mobility Allowance), and generally your income must be very low to qualify.
< www.centrelink.gov.au >

SUBSIDISED TRANSPORT

Some local transport authorities may be able to issue you with a concession bus and train passes if you qualify on mobility grounds, and are unable to drive. There are also taxi transport subsidy schemes around, which are worth investigating. There will generally be a form that your GP or surgeon needs to fill in.

We hope that this chapter has given you some things to think about, and some things to start working on to make your life easier.

Chapter 7
The Emotional Journey

There is no doubt that having hip problems, and the ensuing disability and medical treatments, has more than just a physical impact on your life. In this chapter we hope to shed some light on the emotional aspects. We hope you will find some of this helpful, or at least find some comfort in knowing you're not alone.

It is important that you address any psychological problems you might be having, as they make a difficult situation even *more* difficult, and can impact badly on your self-care, both before and after surgery. The good news is that most emotional problems *can* be improved, and without surgery!

Emotional responses

In Chapter 5 we went through some of the physical aspects of increasing disability and pain. Here we address some of the psychological aspects of these blasted hip problems. There are many areas where your increasing disability and pain can have an impact.

I've experienced the impact all of these emotions at one point or the other. It's been very hard at times. I've especially struggled with how much I've had to lean on friends and family, and hate to be a burden. I'm a very independent person by nature. **Katie, 47 USA**

ANGER

It's a normal reaction to feel angry about your diagnosis and current situation. You may feel angry at yourself, at life, at your hips, at your family, at others who aren't crippled by pain and deformed hips. You may find yourself reacting in unexpected ways, getting annoyed with people's responses to your news, especially if they offer shallow advice, empty platitudes, call you insulting names, or belittle or disbelieve you. You may be feeling let down by your doctors, especially if it's taken a long time to get a diagnosis.

You may be feeling such things as: *Why the hell didn't someone pick it up sooner? Why did my doctors ignore me and my symptoms for so long? Look at those old people walking around perfectly fine, while I'm half their age and limping along on a walking stick! Life isn't fair! Why don't others believe me?*

My husband doubted how much pain I was in. When we would go for walks, especially up hills, he thought I was just bludging and didn't want to exercise. I felt quite resentful about this. **Lea-Anne, 40 Australia**

I have felt angry, angry that I cannot do the things I want to. Angry that I have not only had to contend with the condition and the pain, I also have to battle with the NHS in Wales. Every step has been a nightmare and I have resorted to contacting the Health Minister for Wales and the Medical Director of the Local Health Board. Both have been very nice and helpful, but the process has taken far longer than it needs to. It is hard to know that my paper work was sat on someone's desk for over a month before any forms were filled out. **Sian, 29 UK**

ANXIETY AND DEPRESSION

An increase in anxiety and depression are normal early responses to your diagnosis, and worsening condition. It takes time to adjust. Shock and denial are also common responses!

It's common to go through cycles of acceptance of your condition, and then back into despair. If you find that you're not coming out of these periods of sadness, but are spiralling downwards in to deeper depression, losing interest in life and the things you normally enjoy, and having sleep disturbances (trouble getting to sleep, and early morning waking), it is a good idea to see your GP.

You may be thinking things like: *Life is just too hard, it's all too hopeless. I don't enjoy life any more. I don't have any energy or motivation to change. I'm not coping with my hips getting worse. I'm a burden to everyone. I can't handle this.*

Common triggers of anxiety attacks are: waiting for test results, the diagnosis of a severely debilitating condition, the shock it produces, medical procedures, the side effects of treatment, changes in lifestyle, increase in medical appointments, and increased reliance on therapy and medication.

It is easy when you're dealing with long term and seemingly endless hip problems to get more than just sad and down about it, but actually slide deeper into depression. It is also normal to feel very depressed about three days post-op, so this will generally occur when you're still in hospital. It is quite common to go through a period of depression after surgery, especially when your initial improvement slows down during rehabilitation, as well.

Some possible triggers for depression post-op might be:

◊ Having a caregiver leave (family member has to go back home).

◊ Being a bit more isolated; people tend to visit right after surgery, and then gradually they get busy and lose track of the fact that you still can't drive.

◊ Missing work or school, falling behind.

◊ Overdoing things, trying to do too much too soon.

◊ Sleep deprivation, whether from depression or pain, can also have a drastic effect.

The coping strategies listed later in this chapter may help with some depression; it's also important that you see your doctor (even though you don't see the point, and don't want to do anything anyway). A combination of supportive medication and cognitive behaviour therapy can really make a big difference.

Once you know what your problem is, I would recommend talking to someone who has the same problem, checking out websites, the HipWomen site of course, and becoming as informed as possible. **Lea-Anne, 40 Australia**

FEAR

Fear is one of the most prevalent and common reactions to a DDH diagnosis, especially as treatment is almost always major hip surgery, weeks if not months on crutches, and all manner of tests. It is also frightening to be looking at the months or years of increasing hip problems, and increasing disability, ahead of you. There may well be more hip surgery in your future (a second PAO, a THR, or a revision THR). Major surgery *is* frightening, and it's normal to be scared about the pain, hospitalisation, and possible complications (especially if you have had complications before). If you've had hip surgery in the past, you know all too well what's in front of you. In most cases, however, the months of fear and nervous anticipation of the surgery are often worse than the surgery itself.

You may be thinking things like: *What is going to happen? Will I ever be able to walk? Will I survive the surgery? I'm scared of the pain. Will I get complications? How much pain medication will I need? Will I get addicted to it? Will all this awful surgery even fix me at all? I can't face it. I can't bear it.*

The whole process of having to go through hip surgery again (and the fight to get the right treatment) left me traumatised. I could not sleep and was having nightmares about the surgery. I stepped out of life for a while. On reflection, I was suffering from a kind of post traumatic stress disorder. I had to take time off work before the surgery, I just could not function, which is not like me — really I am quite a robust character normally. **Tina, 41 UK**

GRIEF

It's normal to feel grief at the loss of normal functioning, and increase in pain, and realisation that there is something seriously wrong with your hips, and the prospect of future surgery. The process of grief can be cyclical when you're dealing with long-term disability — you may go through it repeatedly, especially when some aspect of your hip health suddenly gets worse.

You may be thinking things like: *I'm no longer the same person, I've lost myself, I've lost functioning and maybe it's permanent. How can I go on? This can't be happening to me.*

GUILT

Sometimes guilt can become a problem. People can sometimes believe that they are being punished for something by having these hip problems. Some superstitious and religious beliefs can cause guilt.

Your parents may have trouble with guilt, too, whether or not you had a childhood diagnosis and surgery. They may feel dreadfully guilty about having had to leave you in hospital as a baby or young child for traumatic treatments. They may feel remorse if you've had an adult diagnosis of hip dysplasia, and wonder whether they missed seeing something when you were a child, which could have been treated then, and avoided your problems now.

Feeling like a burden on friends and family is a very common experience for hippies, too. As you get disabled, and when you're recovering from surgery, you will need people to help with a myriad of things, from helping you shower to driving you to the hospital for appointments.

You may be thinking things like: *I'm bad, did I cause these hip problems? Should I have been more careful? I must be a bad person to deserved this. I should have done something sooner, then my hips would be better. If I hadn't done 'x', then this wouldn't have happened. I'm being punished. I'm a dreadful burden on all my friends and family, I hate asking for help, they must hate me.*

I tend to ignore pain until it gets really bad, but my hip pain and the way it was going on seemed just very weird. It didn't really impact me much, it would come and go, and I only felt it when I was at home for some reason.

I was going through divorce at the time, so I actually was starting to think it was a curse of some kind from my ex's new girlfriend (that's just to tell you how weird it was — I don't believe in things like curses usually!). Anyway, I only mentioned it to the doctor at my annual check-up and from there all the story with needing a surgery developed, so the surgery news and disability related to surgery was the worst — but not the pain before I had it. **Arpine, 28 USA**

INTERPERSONAL INSECURITY

Your hip problems and increasing disability from DDH has an impact on your nearest and dearest (more on this in Chapter 9). They may not be very patient with you sometimes, or may be annoyed when you have to curtail activities, or cancel plans. You may not feel up to going out as much, and your friends might feel disappointed. This can lead to you withdrawing from your friends, family, and support networks. You can end up feeling rejected, alone, and abandoned.

You may be thinking things like: *How will my family cope? I can't look after my family properly. My relationships are suffering. I don't even have a sex life any more. No one really wants to listen to me complaining, they don't care or understand. My friends don't want to be lumbered with a cripple. I can't get out to my social activities any more.*

I feel like because most of my friends did cheerleading with me, I have lost a lot of my close friends. Even the friends that still hang out with me — it's different. They just don't understand what I'm going through. They always ask me about cheerleading, and act surprised when I get in a bad mood because of it.
Bri, 15 USA

LOSS OF SELF–ESTEEM

Self-esteem is a fragile thing. Our sense of self gets a battering from many directions — how we look, whether we're at a healthy weight or not, how fit we are, how we're performing at work or school, how successful our relationships and friendships are … the list goes on and on! The list of 'shoulds' seems never-ending at times. When you have bad hips, it's easy to feel 'broken' and useless at times, especially when your level of disability keeps increasing (to say nothing about the pain!).

You may be thinking things like: *What's the use? What value do I have, when I can't even bend down and pick up my children, let alone play with them? I'm "damaged goods", limping, weak, in pain, unable to move normally. I've been mangled and deformed by scars, I look horrible. I feel like my body and hips are separate from me. I'm isolated from rest of world, and totally alone. No-one understands me or what I'm going through.*

110

My emotions have always been related to self-esteem when doing gym, jogging, paddling or trekking. I always noticed that my legs didn't move steadily but 'wobbled'. **Alvaro, 38 Spain**

SADNESS

Feeling upset and sad about all this is perfectly reasonable and, really, is to be expected. It would be a little odd if you felt happy about it, after all! (However, it is quite normal to feel relieved and possibly even happy to finally getting a diagnosis, after years of wondering what was wrong.) Hip dysplasia is a difficult condition, it's hard to diagnose, it's hard to treat. It would be surprising if you *didn't* need a good cry now and then!

You may be thinking things like: *I wish I didn't have these hip problems, I don't want to have to deal with them at all. I can't do all the things I love any more. I can't play with my kids. I'm missing out on so much of my life. I feel unattractive, broken, useless, and I don't know how to overcome it. I'm upset about being in pain all the time.*

SELF–DOUBT

Having such an unusual condition as DDH means that you probably don't know anyone else around you who has the condition as well. No doubt, you will get a ton of well-meaning (but misguided) advice from others around you about hip problems — it seems like everyone is an instant expert nowadays! They may cast doubt on your symptoms, and whether things are really *that* bad. If you hear enough of this, you may start believing that you don't really have a problem at all!

You may be thinking things like: *Are my hips really getting worse? Am I just imagining it? Is it really that serious? Is the pain really that bad? Maybe I'm just a pathetic weakling? I can't even find a doctor who knows what they're talking about.*

Sometimes I doubt myself that I am actually in as much pain as I think I am. Also I have lost a lot of confidence and self esteem because I have put on weight. I also felt, and still feel, not in control of my life. **Sian, 29 UK**

UNCERTAINTY

Getting a diagnosis of DDH, and your increasing disability, really throws a serious spanner in the works of your life. When will your surgery be? How long can you hold off surgery before it's really necessary? If you have osteoarthritis in your hip, it can be bearable for quite a long time, but then suddenly deteriorate to a very painful state very quickly. When will that happen? Should you get married, have children, get a job, go on a trip, start your studies? You put your future put on hold while hip problems take increasing precedence. You become "sensitised" to hip pain and hip symptoms, and worry about every little twinge and new symptom.

You may be thinking things like: *Can I commit to anything? Will my hips cope? I have to put my life on hold until after surgery. How long until my hip pain is unbearable and I have to have surgery? Is this new sensation a new problem, or am I over-reacting? How can I cope when I can't walk properly? My plans are postponed, changed, cancelled, or entirely abandoned.*

I was in increasing pain for six years and no-one could work out what was wrong, it was very frustrating as I knew there was something wrong, yet I didn't know who I should see. **Lea-Anne, 40 Australia**

We discuss some coping strategies for all of these emotional responses in the sections below. But firstly, a word about what you're going through:

Validation

You *need* and *deserve* compassion and empathy. Sure, there are plenty of people worse off than yourself — those dealing with acute life-threatening illnesses like cancer, for example — but that doesn't diminish or change the fact that you are living through a *very* difficult time. You have a significant structural problem with your hips that is impacting on your life every single day, with increasing pain and disability. The sorts of hip problems we get from dysplasia tend to be long term and lifelong.

Your experience is valid and shouldn't be dismissed just because "Other people have it worse". Certainly having that sort of awareness may help at times, it *is* good to remember that hip dysplasia *isn't* life-threatening, but don't belittle your own experience!

You still have to live with the pain and disability, and go through pretty awful surgery. It has a real and hard impact on your life. Some hippies need to have multiple hip surgeries over their lifetime, with recurring problems, and complications. There's generally no quick fix.

Coping strategies

Just as there are a host of emotional and psychological impacts from your hip problems, there are a host of coping strategies you can try. We've included a few of the major ones here, but you may like to look for books on coping with disability, chronic conditions, and pain at your local bookshop or library. It may also be a good idea to seek professional help if you think things are getting out of hand, if you might be depressed, or if you're not coping with stress. It is better to start seeking help sooner rather than later, too, as it may take a while to find a practitioner who suits you.

The overall goal here is a peaceful acceptance of your condition, and a pro-active approach to daily life and treatment, without railing against the inevitable and what you cannot change.

I've tried various medications, physical therapy/exercise, TENS unit, acupuncture, water therapy in a heated pool, infrared sauna, meditation, pain cream/patches, glucosamine, vitamin D, massage, biofeedback. Massage and acupuncture were very helpful, but too expensive to do often. The TENS unit enabled me to keep working far longer than I would have been able to on my own. **Katie, 47 USA**

Firstly, don't deny your symptoms. Putting them out of your mind won't make them go away. At some level, you need to accept that you *do* have serious hip problems, and that you need to cope with them, and seek treatment to see an improvement in your quality of life long-term.

Secondly, remember that there is no such thing as 'normal' — *everyone* has something unusual or unique about them, something that makes their life hard, or presents some challenge — maybe they were abused as a child, or have asthma, or diabetes, or have a dysfunctional family situation, or live with an alcoholic partner.

One of our favourite quotations is "The only normal people are those you don't know very well." (Joe Ancis). And a wise man said "Be kind, for everyone you meet is fighting a hard battle." *Everyone* on this planet has challenges to contend with, things that make life a struggle in some way or another. And about 9 million of them have hip dysplasia (0.13% of 7 billion)! You really aren't alone.

Thirdly, try to focus on the bits of you that *do* work, not the small percentage of what doesn't. Maybe you can't run around the playground with your kids, but you *can* play board games with them, read them stories, and help them with their homework. Look at your situation — some things absolutely can't be changed (wishing and positive thinking aren't going to change the structure of your hips), so look at what things you just have to accept, and don't waste energy fighting and railing against them. Look at what things you *can* adapt to, and ways you can do things differently to improve your quality of life (for example, swimming regularly to improve your muscle strength).

Finally, try your best to live in the present moment, day by day. Try not to look too far into the unknown and unknowable future, or second-guess what some doctor or surgeon is going to do or say about your condition. Practically none of the things that we worry about, and are scared of, ever happen. The mindfulness practices mentioned below can help with this.

MENTALLY PREPARING FOR SURGERY

So, you're freaking out about your upcoming surgery (or other medical procedure), but don't want to go to a counsellor (although, honestly, we do recommend you give it at least one try — ask your GP, friends, and family for some recommendations if you don't know where to start). What can you do for yourself at home to help bring your mental anguish into line?

Mindfulness, meditation, and guided imagery are some methods (discussed below) that you might like to try in helping prepare for surgery, both in the months leading up to the date, and in the more anxious days just before and the day of surgery. These can be helpful whether you're just struggling a little with worry and not sleeping well, or if you have more serious anxiety and fear problems. It is a good idea to make an early start on practising these methods, rather than suddenly turning to them in a state of panic a few days before your operation. The list is alphabetical.

EXERCISE

It has long been recognised that exercise is great for your brain, health, and mood. Even if you can't manage much with your dodgey hips, anything you *can* do (a swim, or seated exercises, perhaps) that raises your heartbeat may well help your mood as well. Aerobic exercise has been shown to be as effective as anti-depressants for the treatment of depression for many people[52]. We discuss exercise in more detail in Chapter 5

GOAL–SETTING

Set yourself realistic goals that you can work towards — either goals around your hip health, or work/home/study based goals. Write down your goal, and break it into manageable smaller steps. Your goal might be increasing your fitness level, for example, and going for a swim once a week would be a good step towards reaching that goal.

GUIDED IMAGERY

This is a type of focussed relaxation, using spoken words that you listen to — a 'script' — to create an imagined mental visualisation, generally of peaceful and pleasant images and places, which can help create relaxation and a sense of well-being. The narrator of these scripts might walk you along an imaginary path to a secluded beach for example, to help you reach a deeply relaxed state. For this therapy you either need to see an actual guided imagery therapist, or get a hold of some tapes, CDs, or files from iTunes™. It can help with coping with medical procedures, as you can mentally "Go to your happy place" while they're happening. It may also be of benefit in coping with stress, anger, fear, and anxiety.

It is important to note that the sort of guided imagery that promotes 'healing' of things like cancer (by visualising your body attacking and killing cancer cells) does not and *cannot* work, and is not recommended. We recommend guided imagery for stress reduction and mental calm, but not for "curing" physical problems.

MEDITATION

Meditation is a focused relaxation technique, where you focus on a single word, or object, and try to shut out the flow of thoughts, and create a self-induced state of relaxation and mental silence. Meditation can be done with the help of props such as prayer beads or a smooth pebble, and is often associated with practices such as yoga and tai chi. There are dozens of varieties of meditation, some with religious overtones, and others without. In some forms, a focus on breathing is an important part of the meditation practice. Do some research to find one that suits you. Even five minutes a day of meditation can be beneficial, it doesn't have to take a huge time commitment on your part.

MINDFULNESS

Mindfulness is a type of meditative practice of bringing one's attention to the immediate moment, without judgement, and focusing on everyday things we do, and our current state of existence. For example, to eat mindfully, you bring your close attention to every bite, every flavour, every scent, every sensation as you chew. It was brought to Western attention largely by Dr Jon Kabat-Zinn in the 1970s, from various Buddhist traditions. The practice is generally non-spiritual. Current research does seem to support mindfulness as having a positive effect on stress, pain, and anxiety (although the studies aren't very vigorous, there do appear to be possibly positive results)[53]. There are CDs you can purchase with mindfulness scripts to listen to, we can recommend the ones by Dr Jon Kabat-Zinn and Russ Harris from personal experience, but there are many to choose from. You can get them on iTunes™ as well, and listen to them on your MP3 player.

OVERCOMING PERFECTIONISM

Perfectionism is a personality trait that can get in the way of coping with chronic conditions and disability. When you try to do everything perfectly all the time, you rarely stop to enjoy your achievements, and never give yourself a break. It's a stressful way to live.

When you're unable to do things as well as you used to be able to, because of illness or disability, it can lead to massive frustration. Rather than saying "I'm useless, I can't handle a full time job, and keep the house clean anymore because of my stupid hips," focus on what you *can* do. "I'm managing pretty well with my job despite everything, and at least I can keep the kitchen clean," is a more helpful and relaxed attitude.

Enjoy the process, the *doing*, of activities, rather than the outcome, if you can. There are many books on how to overcome perfectionism if this is something you recognise in yourself; we recommend *Reinventing Your Life: The Breakthrough Program to End Negative Behavior ... and Feel Great Again* by Jeffrey Young and Janet Klosko.

PLEASURE

It is important to build *at least* one pleasurable activity into your day, every day. If you're struggling with depression, feeling like life holds no joy for you, and that everything is useless and bleak, it's *doubly* important, even if none of these things looks interesting to you at all. The list that follows is just for starters. See if you can create a list of your own, even if you have to remember what things you used to enjoy, as you might not feel that you'd enjoy them now. That's the depression talking, not you ...

List of Pleasurable Activities

◊   Call/email/SMS a friend.

◊   Go out and visit a friend if you're mobile.

◊   Invite a friend to visit you at home if getting out is a problem for you.

◊   Call a favourite family member you haven't spoken to recently.

◊   Cook your favorite dish or meal.

117

◊ Cook a recipe you've never tried before.

◊ Go outside and play with your pet.

◊ Do some seated exercises (look online for chair exercise routines).

◊ Go for a swim.

◊ Have a massage.

◊ Have sex with your partner, or masturbate.

◊ Do something nice for someone else.

◊ Sit outside in the sunshine and watch birds, insects, and other animals.

◊ Find something silly to do, like reading comics, or drawing moustaches on the models in a glossy magazine.

◊ Watch a funny movie or TV series.

◊ Play a board or card game with a friend.

◊ Do a 1000 piece jigsaw puzzle.

◊ Do some crosswords or sudokus, or other favourite puzzles.

◊ Go to a free public talk at a local institution.

◊ Go to a bookstore and browse (and maybe buy!).

◊ Borrow some books from your local library or from friends. Try reading a genre you're not used to (poetry, sci-fi, biographies?).

◊ Get involved in public art or treasure hunt projects online (PostSecret, mail art, Letterboxing, BookCrossing, Postcrossing, ToyVoyagers, and The Toy Society are just a few!).

◊ Sing along loudly with the radio or your favourite CD.

◊ Draw or paint a picture.

◊ Write a poem.

◊ Write in your diary or journal.

◊ Write a 'paper and ink' letter to a friend.

◊ Make a collage using pictures from glossy magazines, a glue stick, and scissors.

◊ Get outside, tidy up, weed the flower beds, do some pruning, or work in your vegetable or flower garden. Use long-handled tools if you have trouble bending over or kneeling.

◊ Get creative: knit, crochet, whittle wood, embroider, sew, build model airplanes, do origami, or whatever you enjoy — or learn how —there are tons of free how-to videos and project instructions online, as well as online and 'real life' craft groups you can join.

◊ Do an online course in a subject you've always wanted to learn more about, such as a new language.

◊ Get involved with an actitity in your local community that you're passionate about.

◊ Help your kids do a craft activity or learn a new skill.

◊ Make a list of things you're looking forward to doing post-recovery.

POSITIVE FOCUS

Try to see positives in your situation, especially with the outcome of surgery — you will be *so* much better after surgery, able to bathe your kids, walk the dog, get back to your favourite sports, studies, and activities, and so on. Keep an eye on the end goal — happier hips! The good news is that this level of disability and pain is (hopefully) unlikely to be life-long, and surgery is a very effective treatment for most patients.

PROBLEM SOLVING

Figure out ways that you can cope better with various aspects of your disability that *can* be changed, there are lots of tips in this book to help you manage better! Don't dwell on what it's impossible at the moment, but look at what you *can* do, with some adaptations where necessary. Maybe move the saucepans from a low cupboard up to a higher one, for example, if bending down low is too painful.

RELAXATION TECHNIQUES

Relaxation is one way to help with the anxiety and fear associated with upcoming surgery and distress at your diagnosis, and depression from chronic pain.

Relaxation techniques can help reduce muscle tension and chronic pain, and slow your heart and breathing rate. It is important to *practice* relaxation techniques. While they may seem ineffective at first, you will get better at them over time. Many people dismiss relaxation before giving it a good try. It is a skill that needs to be learned and

> HIP TIP: Listening to peaceful classical music like Vivaldi and Pachabel has been shown to help people relax more quickly after a stressful event (such as an unpleasant medical procedure), more so than with jazz or pop music[54].

practiced to be effective. Set aside a couple of times a day when you can practice relaxation. You will gradually get better at noticing how tense your muscles are, and how best to physically relax them. You can find books and information on the details of these techniques in your local library, bookshops, and online. They can provide you with greater detail on how to do each of the methods listed below.

Some basic relaxation methods and relaxing things to do include:

◊ Deep breathing

◊ Progressive muscle relaxation

◊ Guided imagery (see above)

◊ Massage

◊ Hot or cold showers

◊ Rest days

◊ Listening to music

◊ Watching a movie

◊ Exercise

Here are some resources you might like to look for:

◊ *The Relaxation and Stress Reduction Workbook* by Martha Davis, Elizabeth Eshelman and Matthew McKay

◊ *Full Catastrophe Living* by Jon Kabat-Zinn

◊ *Mindfulness for Beginners* (audio) by Jon Kabat-Zinn

◊ *Mindfulness Skills* (audio) by Russ Harris

SELF–MANAGEMENT

Take charge of your case. Complement the efforts of your doctors, surgeon, therapists, and carers. Keep records of your medical appointments, and be an educated patient about your condition (this book will help!). Keep symptom diaries if you find them useful, or if your doctor asks for them. But also don't allow your hips and medical appointments to become the obsessive centre of your life, there's much more to life than hip dysplasia, after all!

When you have to deal with things that really can't be controlled or avoided, like constant pain, medical tests, and procedures, you may find some emotion-focused coping techniques helpful in calming your mind. These include such techniques as mini-meditation breaks, deep breathing, keeping things in perspective, looking after your sense of humour, and cultivating optimism.

We know it's a cliché, but do your best to accept the things you can't change, and change the things you can. It really does hold true when it comes to chronic health conditions like hip dysplasia.

Remind yourself that having hip dysplasia is not your fault, nor your parents' fault — these things just happen. It is outside of your control, but you still have responsibility to look after yourself, do your physio and treatment rehab, and care for yourself as best as you can, to have control over the situation. Don't just give up and abdicate responsibility to your doctors. This will only add to your sense of helplessness.

It can be easy to get into a 'catastrophising' state of mind, where everything becomes a huge problem and drama, destined for the worst outcomes. Try placing possible outcomes and events on a Catastrophe Scale, where 100 is the worst possible event, like the death of a loved one or a life-threatening illness, 90 is a serious event like a car accident, 50 is 'medium bad' events, 10 is a little bit bad, and so on. Having to assess your fears about surgery and disability like this, and place them on this sort of a scale can help to put things into perspective.

SENSE OF HOPE

It's important to maintain hope. There are always new therapies being developed, new treatments being trialled, new research discovering more about hip dysplasia, better hip prostheses being developed, better screening procedures being set up.

You can also have hope that your condition will improve quite significantly with appropriate medical treatment, even to the point where your hips won't be a daily problem for you at all. This does happen for many people, and has happened for us, so it is not a false hope at all!

I liked reading blogs and posts and updates from people who've had successful surgeries and live active lives, as this gave me much hope that I will be at that point as well. That's partly the reason why I try to send a HipWomen note when I get my check-ups, even though I don't have much to report — remembering how those positive notes helped me, I am hoping to help others showing that even with complications, there can be success, which gives a lot!
Arpine, 28 USA

SURGERY SCRIPTS

There are some specialised 'surgery preparation' relaxation CDs. Many of these can be purchased or downloaded online, and it's generally possible to hear previews of them to make sure you like the voice (there's no point getting one which is going to make your toes curl every time you hear the narrator's voice!).

The positive thinking myth

You know what? It's okay to be sad and angry about your situation. It *is* crap, there's no two ways about it. You honestly don't have to be 'upbeat' and cheerful all the time, putting on a brave face.

There's a myth about the power of positive thinking which is just about accepted knowledge around the world. However, *no* research — and plenty has been done — shows that thinking positively will change your health outcomes, whether it's hip dysplasia, heart disease, or cancer. So don't worry that you're somehow harming your recovery or condition if you're feeling pissed off, sad, or depressed about it all.

Where a positive attitude *can* help is with how engaged you are in the preparation for surgery, your time in hospital, and your recovery. If you are feeling upbeat, you're more likely to take your medication as prescribed, do your physio exercises, prepare for hospital without dreading it quite so much, follow your rehabilitation program, look after yourself, and be much easier to live with. But while these can lead to a better outcome, because you're more involved in and responsible for your own health, simply having a sunny disposition isn't going to ease any aches or pains.

So rest assured, you aren't harming your recovery or condition if you're having the occasional down or grumpy day. All research into this has *failed* to show that having relentlessly 'positive thoughts' staves off illness or speeds up recovery[55].

Creative approaches

WRITING THINGS DOWN

Journals

A classic journal is a thing of beauty, and a well-established tool for reducing stress. It allows you to express yourself in a private place, and leave your troubles, fears, and anger on the page rather than in your head. It's nice to have an excuse to buy a beautifully bound blank book and maybe even an elegant fountain pen especially for the job — but a plain exercise book covered with a bit of wrapping paper you like and a ballpoint pen will do the job perfectly well if cost is an issue. It really does help to write stuff down, and has been clinically proven to reduce stress and even improve health outcomes[56,57].

If you can, make time to write in your journal for a few minutes every day. Don't worry about your spelling or sentence construction, it's just a safe place where you can get your thoughts down, however that may be.

Some days you might write several pages, others it might just be a couple of words. Sometimes it might be a picture, or some lyrics from a song that means something to you, a page full of swear words, or the beginnings of a poem that is forming in your mind …

Blogs

Quite a few HipWomen have blogs (short for "we**b log**"). A blog is an online diary where you can post photos and diary entries, and other people can leave comments on what you write, if they so desire. Blogging is generally free; WordPress, Blogger, LiveJournal, and Vox are just a few of the main blog providers. When setting up a blog, you can decide on what level of privacy you want — visible to everyone, restricted access to a set of friends who get see it by using a password, or completely private just for yourself.

Running a blog can be a great way to chart your hip journey, and share your experiences and even x-rays with others who have been or are going through the same thing. Many of us have found a great deal of help from reading others' blogs, as well as solace in writing about our own experiences in this way. There is more information about blogs, and a list of "hip dysplasia blogs" in Chapter 17.

My first port of call for information was Hobbling Helly's blog < hobblinghelly.wordpress.com >. This led me to all sorts of different websites telling me all about hip dysplasia, but most importantly it led me to the HipWomen Yahoo Group! **Annick, 47 UK**

I really enjoyed reading Denise's blog < jejunesplace.blogspot.com > and found it to be very helpful on my journey. Plus, it was always a treat to see that cute little puppy Petal! **Ina, 41 USA**

Figure 7.1: Petal the cute little chihuahua who knows nothing about hip dysplasia.

124

ART THERAPY

If words fail you, it's also very helpful to do drawings and paintings, or even sculpture. These aren't created with the goal of making works of art for display, but are a way of expressing yourself and your emotions, where the act of creation is therapeutic. This can be a safe way to process traumatic experiences and turbulent emotions, especially if you don't want to write things down. You can incorporate any imagery that comes to you in your drawings and paintings.

Gather a few art materials — this can be as simple and inexpensive as a box of coloured pencils or some coloured ballpoint pens, or as elaborate as a range of watercolour, acrylic, or oil paints, and brushes. Pastels, crayons, and ink pens are other options you may like to try. The main thing is to find a medium you like using, and that you won't worry about using up (frightfully expensive paints might hamper your style!).

The only other thing you need is something to create your art on — anything with a blank surface will do, from the back of notes from school and student-quality drawing pads, to bound books of watercolour paper and stretched canvas.

To start, find yourself a quiet place and time, when you won't be disturbed, if possible. Take the phone off the hook, turn off your mobile phone. Put on some music that suits your mood. Have your paper and art materials to hand, and a refreshing drink if you like.

Think for a few minutes about how you're feeling about your hips and surgery. Are you scared? Are you depressed? Angry? Feeling isolated? Choose a colour that represents your mood, and start to make marks on the page — depression might be a dark green scribble for you, anger might be black and red zigzags — build up your picture gradually.

You can add words if you like, or even cut pictures out of a magazine and stick them onto your artwork. Cut, draw, scrape, paint, tear. The *activity* of creating is the main focus here, not the end product. Everyone will create something different, so don't feel bound by any conventions. Don't worry about how the end artwork looks, that's not the point of the exercise, no-one else is going to see it (unless you choose to show them). You can even throw it out or burn it afterwards, if you want to.

A web search on "art therapy activities" will bring up a host of ideas to get you started. The blog < www.arttherapyblog.com > is a good place to start.

Here are some ideas you might like to try:

◊ Sketch some abstract doodles and patterns, let your mind roam and daydream.

◊ Paint abstract shapes and figures. Don't try to create any representational images (e.g. people, buildings).

◊ Paint a self-portrait. It can represent where you are now, what you would like to be, you as a child, whatever you like. Remember, this isn't going to be shown to anyone else, and doesn't need to look like you, you don't need to use a mirror if you don't want to. It can be as abstract as you like.

◊ Keep a drawing journal. This can be in the same book as your written journal, or in a special sketch book of its own.

Clinical art therapy, where you attend sessions with a trained art therapy counsellor, can be especially helpful in cases of trauma or long-term problems such as depression and anxiety. Look in the Yellow Pages or online to find qualified therapists in your area.

Scars

Scars can be a mental hurdle for some. Childhood scars can be thick and ugly, and may have caused you grief at the pool or beach — children can be notoriously cruel, and teasing and bullying from looking different is a wide-spread problem for many kids, not just those with hip dysplasia.

"Modern" scars seem to be much less noticeable, from improvements in surgical technique, suturing methods, wound care, scar treatments, and an appreciation from the surgeons of positioning of scars to get a good cosmetic result. So more recent surgery is less likely to result in really ugly scarring.

If you are *really* unhappy with your scars, there are plastic surgeons who specialise in treating them. Despite the advertising claims, once scars have healed, special skin creams won't make a significant difference to their appearance. Firm massage along the scar can help with releasing scars that have become stuck to underlying tissue.

Many HipWomen wear their scars proudly as badges of honour or valour — they're our "battle scars", visible proof of what traumas we've suffered, what surgical battles we've waged, visible proof of what was wrong inside. Quite a few people with invisible illnesses (think of endometriosis, for example) would be quite glad of some visible manifestation of their internal woes, it certainly helps people to realise something is actually wrong.

I think scars are pretty cool. Reminds everyone how tough we are to have endured what we have. If I'm going to go through two PAOs and a THR, then I want a battle wound to show for it! I have symmetrical scars down the front of each hip, they go nicely with my Caesarian scar across the top. And an appendectomy scar, a screw removal scar, and some stretchmarks for good measure! I'm collecting the whole set! **Sandra, 39 Australia**

I look back now and find it like a dream. As my husband said to me, "If it wasn't for the faint scars on either hip, you would never know anything had been done". And it's true, although if you could look inside me, the evidence is there, not just physically, but mentally. I am different to the person I was two years ago. **Annick, 47 UK**

Childhood experiences

This topic is often neglected, but a very real and important one. If you were diagnosed with hip dysplasia as a baby or young child, there are many major benefits for your long-term "hip health", as the problem was picked up early, and hopefully you received effective treatment. Of course, if you were a baby at the time, you might not remember any of this at all. Babies are resilient, and hopefully the early treatments were successful so you didn't have to go through anything too traumatic as a child.

However, if you were a toddler or older, you may well have some memories of your hospital stays. You will probably have also experienced a dreadful lot of tests, splints, medications, plaster casts, braces, traction, hospitalisation, and surgery at a very young age. You were possibly the "weird kid" at school, in plaster, or on crutches, possibly limping, unable to run properly, not very good at sports, and missing lessons for doctor and hospital visits.

All of this takes a toll. There's only so much you can understand of what's happening to you when you're a toddler, and only so much you can bear as a school-aged child. No matter how much your parents supported and stayed by you (and it was an awful experience for them, too, having to take you in to have all these things done to you), in the end you still had to suffer often dreadful treatments by yourself. No matter how kind the nurses and doctors were to you, they still hurt you. Although the intent was always to help you, you *did* suffer from unusual hospital experiences (different from what your peers were going through at that time in your life), and painful physical procedures which you didn't understand, and couldn't avoid.

When I was a child I noticed that my gait was abnormal, not much, but enough to makes me feel different. I remembered a friend of mine telling me that I walked like a bell. **Alvaro, 38 Spain**

I remember being so frustrated at having to spend all summer in full body plaster instead of going swimming. Because my last major hip surgery as a child had been experimental, I was 'Exhibit A' in a couple of medical lectures. I remember lying on a table at the front of the lecture theatre while my surgeon lectured, and medical students looked at my hip and leg. **Denise, 47 Australia**

TREATMENTS FOR DDH IN CHILDREN

This is a brief list of common treatments for hip dysplasia in children, some — or all — of which you may have had.

◊ **Closed reduction**
The surgeon manipulates the hip under the guidance of x-ray, positioning it correctly. The hips are then plastered to maintain this position.

◊ **Open reduction**
The hip socket is opened surgically, its position corrected, and then the hips are plastered to keep them in position.

128

◊ **Osteotomies**
The surgical cutting and repositioning of bones. These are usually done in slightly older children, or in those where the diagnosis was missed as a baby, or where other treatments have not worked.

◊ **Pavlik Harness**
A contraption made from canvas straps, Velcro, and buckles that holds the legs apart and in a particular position, generally a "W" shape, which can help the hip socket develop correctly. It is just used with babies.

◊ **Hip spica cast**
A cast which covers the trunk and both legs, generally only used for hip dysplasia patients. There is often a bar running across the knees, to maintain the 'frog leg' position. Plaster typically goes from mid-chest to below the knees, or even to the toes. There is a hole at the groin for toileting and nappies.

◊ **Hip abduction orthosis (brace)**
A brace is sometimes used instead of the Pavlik Harness, or when the Pavlik isn't working. In some cases it can be used instead of a spica cast, if the hip easily goes into the socket and the doctor feels the parents will comply with brace wear. It is also common for a child to have a brace right after the spica cast comes off to keep the hips stable, and also because loss of muscle tone occurs while the child is in the cast. It takes time to regain the strength that was lost.

Some treatments that are no longer routine for hip dysplasia include:

◊ **Tendonectomy**
Cutting of a tendon in the hip, generally not carried out in humans any more (but is used by vets!).

◊ **Traction**
The application of a sustained pull on a limb using weights in order to correct a deformity. It can also gently stretch muscles and tendons around the hip joint, either before or after surgery. In some extreme cases, traction can involve metal bars inserted through the leg itself, and the weights pulling from this point. (Traction is still used during procedures such as hip arthroscopy, though.)

CHILDREN IN HOSPITAL

Hospital in the "bad old days" (pre-1980s) often wasn't a very child-friendly environment. Child psychology was only just developing as a field in the 1960s. Even until the 1970s, doctors and nurses were taught that babies didn't feel pain, supposedly because of the immaturity of their brains and nervous systems. This is now known to be entirely incorrect, and that babies in fact feel *more* pain than adults for the same procedure. Doctors were often anxious about administering anaesthetics to young babies and children, as they weren't sure if they were safe for them.

In fact, suffering severe and ongoing pain as a baby can have long term consequences, even changing one's physiological and emotional response to pain, immune function, neurophysiology, attitudes to medical procedures and avoidance of health care as an adult[58]. Nowadays, there is much more care given to ensure that babies and children are well covered for pain relief!

The "old-time" hospital environment was really not good for kids. Parents were oftem not encouraged to visit, as the children just got upset when Mum and Dad left, and that made the nurses' jobs harder and they felt parents got in the way of their duty to help sick children get better[59]. In older times there may have only been a couple of days a week when parents were allowed to visit. Parents certainly couldn't stay overnight, unlike the more enlightened situation in hospitals today. Children were alone at night, often in the dark, and nurses could be impatient if their young charges cried or carried on. Discipline was strict. Practices such as force feeding happened to children who refused to eat[60].

Please keep in mind, though, we are not saying that every child had bad or traumatic experiences; some hospitals were more enlightened than others — one HipWoman says that when she was in hospital as an infant in 1962, her mum was allowed to come in every day and breastfeed her. Some children certainly received good, compassionate care, and don't have any trauma associated with their childhood hip surgeries. This section is mainly written for those of us who had a rough time of it.

I don't remember a lot but I know my mum could stay with me throughout due to how little I was! The main thing I remember is a dolls' house at the hospital in Essex, it was the most beautiful dolls' house I have/had ever seen, I remember seeing that at all my check ups up til I was 5 years old! **Rhianna, 28 UK**

I was hospitalized at least twice in the 60's. I had a closed reduction first in 1965, followed by an open reduction and osteotomy ("shelf procedure") in 1966. I was diagnosed at age 4½, and the surgeries/hospitalizations/casts/braces lasted for about two years.

Parents not being allowed to stay was horrible — I hated it (however, the chocolate milk and graham crackers were great). Pain control wasn't good at all. Being told to "be good" and not cry was ridiculous. My surgeon casted my doll, which was great (I still have her, although her plaster's wearing a bit thin. The poor girl has been through a lot). I hated not being able to go play with my friends. I hated not starting school on time (no kinder, and didn't attend class until May of first grade.) The kid across the street made fun of my limp, and I had to ride the "special bus" to school May and June of first grade. Not so great for the self-esteem. **Sally, 51 USA**

Many hospitals had rigid standard procedures, it was a regimented and hierarchical system; children may have had things like daily finger-prick blood tests, whether they needed anything tested or not. It was common for patients and their parents *not* to be kept well-informed about what was happening or being planned. Nurses were under pressure to run a tight ship, and used disciplinary measures to keep order in the wards. Sometimes restrictions were invented, simply to keep order. In the worst situations, the atmosphere could border on the military and even be unfriendly[59].

My first vivid hospitalization stay was when I was 11 years old and was admitted for traction. I recall the weight feeling heavy and felt the pressure on my hips. It felt very uncomfortable because I could not move and was stationed in my bed for two weeks. I bathed and used the restroom lying in bed. It was not a pleasant feeling. I recall feelings of isolation and fear. Many of my fears included the unknown and blood drawn twice a day. I did not understand why they had to take my blood, which was not easy with my tiny veins. I was poked several times before they could get any blood. **Cathy, 39 USA**

I spent ages 9 –12 in and out of the hospital, in the 1970s. The surgeries and physical therapy had been very painful and not much fun (of course). I blamed God for all of it.

My mom came to visit often but I always missed her. We were very close. I became very depressed and *angry* as a child. I had several therapists try to work with me, but it became a very ugly scene. My roommate and I became the trouble makers of the ward. Leaving the hospital, going on the roof, and so on. I was so angry and frustrated. I didn't understand why this was happening to me. It was my way of dealing with it. Getting into mischief. Rage against it! I wanted to be outside running (something that I would never do again), but couldn't except in my dreams. **Ina, 41 USA**

Eating was only thing I had control over in hospital, when I was 3. I refused to eat. My parents came in to visit one day to discover two nurses holding me down on the floor, force feeding me. This was in 1967, in a major children's hospital in Melbourne. I also remember being fully conscious on the operating table many times (with no pre-surgery medication to make me drowsy), and the black rubber oxygen mask being held forcibly over my face. Since then, the smell of black rubber makes me physically ill. I remember being scared and alone in the dark every night; I needed to sleep with a night light on well into my adult years, and have many PTSD symptoms that persist at a low level, but 'flare up' when I have to deal with surgery again. **Denise, 47 Australia**

I was first hospitalised at 6 weeks, then again at 1 year, 2 ½, then again at age 4. I don't remember much — but I do have memories of the hospital — I know it was more traumatic for my parents. They had to drive to San Francisco daily to see me, and my mom always regrets having to leave me there alone. In fact, I am going in for a THR in October and she is adamant about staying with me this time! I do remember crying when my parents left. I also remember this nurse who was so incredible — I just adored her. **Teri 38 USA**

The loss of control was scary for me. The separation from my mom each evening during my hospitalization stay was always an unpleasant and scary feeling for me. I was frightened of painful procedures, and my mom was not available to comfort me and tell me that everything was going to be okay. Being left in an unfamiliar environment with strangers was very stressful for me. When you're alone as a child your imagination runs wild. This is why it is so crucial that children are told the truth, so they have an idea of what to expect, which can help the level of stress reduce.

I know now that my mother experienced shame and guilt as she felt responsible for my condition. Fortunately, my mother provided me with a stable, secure, and loving environment. She was very predictable and I knew when she was coming and going. She provided me with a secure daily routine, and this gave me confidence and security. She reassured me that she was coming back the next day, and she did! **Cathy, 39 USA**

POST-TRAUMATIC STRESS DISORDER FROM CHILDHOOD SURGERY

Post-traumatic stress disorder (PTSD) is a recognised condition, often seen in people who have survived natural disasters, war, horrific accidents, and other traumatic events. It is known that an event is more likely to be traumatic if it occurs during childhood, including surgery in the first three years of life[61].

Medical procedures in children can *sometimes* lead to PTSD — they have suffered physical harm, after all, even if it was treatment to help them, performed by kindly doctors and nurses. The child may still feel as *if* they have been tortured. Add to this the child's sense of being in an unsafe and unpredictable environment in hospital, separation from their parents, and invasive medical procedures, it's hardly surprising that some children have psychological shock that remains and develops into PTSD.

Some signs of PTSD can include:

◊ Avoidance of things, events, thoughts that remind you of your experiences (hospitals, medical appointments, medical shows on TV, images of surgery, needle phobia).

◊ Loss of memories around traumatic events (which can be hard to remember anyway if you were very young when you had surgery).

◊ Nightmares about surgery.

◊ Unable to speak about the traumatic events or your past medical history without crying or getting upset.

◊ Physical stress when reminded of the trauma (sweating, faster heart rate and so on).

◊ Sense of shortened future (you don't think you'll survive further surgery, or that you're not going to live to old age).

133

◊ Detachment, emotional numbing, some distance from people including family.

◊ Hypervigilance (jumpy), startling easily at sudden noises.

◊ Irritability, sleep problems.

Symptoms can develop well after the event, even years afterwards, and persist. They can wax and wane — so they may not be too problematic for most of the time, but having to face more surgery as an adult can bring everything to the fore again.

PTSD is more likely to develop if you were harmed directly (as opposed to just witnessing something), which is the case if you had traumatic childhood surgery. Often there will be "triggers" that set you off — certain smells or tastes that take you back to your hospital experiences, or certain procedures or medical equipment that do the same thing.

One thing that's stayed with me are particular tastes/smells. I *hate* the brand Aquafresh toothpaste — there's something in it that is *just like* something in the rubber mask used for anesthesia when I was a kid. I also hate hospital scrambled eggs, and *will not* eat apple jelly! **Sally, 51 USA**

Treatment for PTSD

There are various treatments for PTSD that are available, some more well-established than others. Cognitive Behaviour Therapy (CBT) and Exposure Therapy can be of some help. Eye Movement Desensitization and Reprocessing (EMDR) is one methodology which is rather controversial, but some people swear by it. It involves reliving traumatic experiences, with the guidance of a trained therapist, while tracking a swiftly moving stick with your eyes. There is no explanation for how EMDR works, though, and not very thorough research into the method.

Medication can be very helpful with PTSD, especially in the time of most trauma leading up to your surgery and hospitalisation. Taking something like diazepam in the week before surgery can work wonders, and is not harmful or addictive over such a short time. Talk to your GP about this. You don't have to have an official diagnosis of PTSD to get the help you need; if you're really freaked out by your upcoming surgery, and having serious problems with anxiety and fear, then you need some help, and can get it.

It is really important that you inform all your medical team and nurses on hospital forms, and your patient notes — make sure they know you have PTSD or trauma problems! Get it put in your notes! If your medical team know about it, they will do their best to support you and reduce exposure to triggers.

While you may not be able to completely overcome your PTSD, if it's from childhood surgery, it's more likely that it will only really present a problem for you in the period around your adult hip surgery and hospitalisation. You may like to see a therapist about dealing with it during the year before surgery, as it isn't anything that can be improved in a hurry. If you have only a short lead time before surgery, look at the "quick temporary fixes" like medication to help you cope. They really can work wonders.

After having a lot of childhood surgeries, it can be a huge challenge having to front up for more hip surgery as an adult. Our childhood experiences can have a range of effects.

I have always had a slight fear of hospitals. I am getting anxious about my upcoming surgery. I remember the gas they gave and once when I was little and at the San Francisco Zoo we walked into the bird exhibit, and for some reason it made me think of the same gas smell from surgery and I ran out of there so quick! Every once in a while I smell something that reminds me of that and I get a feeling of almost panic.

I have not been diagnosed with PTSD, but this upcoming surgery is causing some trauma for me now. The only surgery I have had as an adult was on my foot and I was an outpatient so it wasn't that bad. This upcoming hip surgery is something I am excited about but also terrified of. **Teri, 38 USA**

It brought up a lot of memories. Memories that I haven't wanted to think about, and haven't thought about since being in the hospital as a child. I always felt ashamed that I had dysplasia and never spoke about it. Now it was in my face, and I was having to explain to friends what was going on. It felt like a cleansing process ... It felt good to let the world know. I have Hip Dysplasia, Hear Me Roar — LOL! **Ina, 41 USA**

My childhood experiences have definitely left an impact on how I cope with my future surgeries. For one, I know what to expect in terms of pain, medications, side effects, recovery, and walking with assistance walking devices. Taking action in my care helped me feel stronger and better equipped to deal with lots of life's trial and tribulations. **Cathy, 39 USA**

I suppose my childhood surgeries have had an effect on how I approach my adult hip surgeries. I'm pretty stoic and matter of fact. **Sally, 51 USA**

Living with hip dysplasia presented me and my family with many challenges. My childhood memories triggered unpleasant memories for my upcoming adult surgeries. You learn to adapt at an early age with limitations and chronic pain. Due to this, it is natural that I have a history of depression and anxiety. Losing control at a very young age has led to anxiety and fears. I'm currently seeking professional help for my anxiety disorder. I suffer from palpitations, insomnia, muscle tension, anger, and agitation. Supportive counselling, being part of a support group, and the support from family members and friends has helped me get through difficult moments in my life due to my orthopaedic problems. **Cathy, 39 USA**

I really didn't want to face up to my adult hip surgery. I was effectively putting my fingers in my ears and shouting "La! La! La! I can't hear you!" for years, as my hip's condition deteriorated. My lovely physiotherapist had to coax me for months to get me to see my sports physician, and then a surgeon, about the deteriorating condition of my left hip.

Actually booking in for the surgery was really hard. I couldn't talk about it without crying, and had surgery nightmares. I was convinced I was going to die on the operating table. I had nearly bled to death on the operating table when I was about 3, and have a scar on my ankle where they did an emergency blood transfusion. I was so petrified that I had to be on Valium for the week before surgery (which worked a treat, I must say!). I did go to some counselling, but didn't find it helpful.

I had the academic scores to get into medicine after high school, and while I had (and still have) an interest in medical science, I couldn't face ever having to do anything invasive to people (putting in an IV, taking blood, using a scalpel, doing surgery), so I guess my childhood experiences affected my possible career path!

I think they have also made me pretty pragmatic and stoic about most medical stuff in general. I don't fuss too much about things being done to me (although I don't like them of course), I don't complain too much, and just get on with it. I am trying to use 'exposure therapy' to help ease some of my trauma problems — I force myself watch when I have blood tests taken (I used to get faint), and can now watch most of the procedure. (I have an auto-immune disease, so I have frequent blood tests.) **Denise, 47 Australia**

We hope that this chapter has given you some insights into the psychological aspects of dealing with hip dysplasia, and a few ideas of ways to help you cope.

Chapter 8
Young Adults with DDH

While hip dysplasia is often picked up in infants, and DDH tends to develop later in life, there are those patients who find themselves with increasing hip problems, and a DDH diagnosis and surgery in their teens and early 20s. This is not a good time to be lumbered with all these problems that make life much harder. They make you stand out, and make it difficult to fit in with study commitments, work responsibilities, family expectations, community involvement, and social activities. We've asked some young women with dysplasia to share their experiences and insights.

Emily is 18, and lives in the United States:

One of my pet peeves of having hip dysplasia and the PAO surgery at my age is my new set of physical limitations. I am such an active person and it is really disappointing that I have to sit out on some activities like skiing, hiking, running, and water polo which I played in high school. Just in general I have to be more cautious about what I do. I also grew an extra bone during recovery, which is supposed to be taken out when my screws are removed, and that limits my hip mobility and makes it hard to do some things. First going in to my surgery I thought that by recovering 100% I would be back to my normal self. But now, at 8 months post-op, I realize that I might have to deal with limitations my whole life. This is just my case, and I am sure that other people have different experiences.

I do not regret getting the surgery this early. I am lucky that my doctor caught the hip dysplasia this early. I recovered very quickly, I know that a big reason for that was because I am so young.

One thing that did annoy me a little was that during recovering, when I was using a cane, I had a lot of people staring at me. I eventually learned to ignore it because I think that people were just surprised to see such a young girl with a cane. It was especially hard because I went into my first year of college (university) with a cane, a time when I was trying to make a lot of new friends, and I felt that I really stood out, and I was a little self conscious about it.

The thing that I am most thankful for having my surgery at an early age is that I had my mom close by to help me with the recovery. I could not have done anything without her — she helped me take showers, go to the bathroom, and get dressed. I am so thankful to have her there with me during the process, even if it meant giving up all my pride and independence. I cannot imaging going through this without her.

Hannah is 17 and lives in the UK. She had a THR at age 17. She runs the Hip Dysplasia Awareness and the HappyHips websites for teens:

< www.hdawareness.moonfruit.co.uk > and < happyhips.webs.com >

When I started senior school I had never thought about my hip much before. I just knew I found certain sports like cross country hard and painful, so had to sit out. My friends always said that I was just faking it, as I walk perfectly fine. Last year my friends saw me in a way I wish no one ever had to. They realised what was really wrong.

When I was first told that I needed surgery I cried the whole way home, but then I saw my friends and teachers, they told me I would be OK, and I felt better. They were really supportive. Everyone constantly tells you that you will be OK, and sometimes it doesn't help at all, as it just makes you keep on wondering when this time will come, and you realise how little they understand.

The thing that I learnt from this experience is to never hold it all in, just to let it go, they don't think any less of you if you cry — they realise you're struggling so give you extra support. I was very good at trying to hold it in, as I didn't want my friends to see how much I was struggling, but I soon realised that they were not supporting me as much as I needed. When I spoke to them about it they told me that they hadn't realised how much I was really struggling. One day I just told them all that I needed more support, and they did it straight away, as they could tell how hard it was for me.

140

Each time I explained it, it got easier and I got more confident. Sometimes you get strange looks and people ask how you hurt your leg, but now I just turn round and say 'I haven't done anything, I was born with a hip condition'. They look shocked and feel bad, as they are only trying to be supportive. My school are great — they support me every second of every school day! I know they are always there for me.

The best thing I ever did was to try to contact other teenagers in the same situation. I have made lifelong friends through this. I will never be able to thank all of the people on the forums that I am a member of, and the charities which I contacted during this time. I also hope that in the future my lifelong journey (especially my THR) will allow me to support others in a similar situation.

Rhianna is 28, and lives in the UK. She had a THR at age 22:

I was diagnosed from birth and I had my hip pinned at 6 months. I then had both legs in plaster until the pins came out about six months later, after that I was baby Robocop in what was then called a diverigator, but is now a little more subtle harness! I also always hated it at school or parties where you had to bring in a baby photo, as all mine were with me with my legs wide open either by plaster cast or a metal bar separating them! Never a good look when you're 14, and the boy you have a crush on is in your class, and he asks to have a look!

Because I have always been aware of my condition growing up, I didn't have any problems until I reached secondary school. I have a very large scar on my hip from childhood surgeries, and a 1 ½ inch leg length discrepancy — I had my shoes built up to compensate for this. With my close friends at school (four of whom are still my best mates now), I told them all that I knew from Mum and my consultants about my surgery. My mum had x-rays of my hips as a child right through my teen check ups, so I showed them!

I have never been afraid of needles or surgeries or anaesthetics. I was always the first in line at school for injections showing all the so-called cool kids who were quivering in the corner that I was like "What's the big deal, I'll go first", which made me look tough and cool for about an hour!

And I also did my science book report on it, going into details of surgeries and my x-rays, and grossed a few of them out, including some of the teachers I'm sure! I have never ever been ashamed of my scars or my medical history. I have had such an interest in medicine since I can remember, so I had learned all about my condition and surgeries. I've never felt embarrassed by it, as I always felt quite in control of things, and confident and quick witted enough to come back with something funnier if anyone made any comments about my scars or limp. All of this is an absolute credit to my parents and family for being as open and honest as they were whenever I asked any questions.

It only really bothered me when my hip gave up on me at 20 years old! I had just met my now husband and I found it hard when we were all going out for the night, the girls getting ready together like we do, and my friends were sharing shoes and all had beautiful killer heels. I was so jealous, as all I could cope with was a flat pair of ballet pump style shoes.

Bri is 15 and lives in the United States. She had a PAO at age 15:

Not only does hip dysplasia physically hurt me, it emotionally hurts as well. These past few months I have felt so alone. I feel like no matter which friends I vent to, they won't understand half of what I'm dealing with.

But after spending hours on the internet, I realized that I'm *not* alone. There are a few blogs with teens sharing their experiences. The blogs I have read have comforted me. They made me realize that others know exactly what I'm going through. I'd strongly recommend checking them out.

Also, after being diagnosed with hip dysplasia, I have had to quit cheerleading. This was probably the hardest thing on me. All my friends talk about it, and I feel so left out. Nothing makes me more sad than watching my former team compete without me. I worked so hard for so long, and I wish I hadn't taken it all for granted.

Bri's mother Cynthia also writes:

Since writing this, Bri has come up with a plan. She's always been an excellent student. In the time she's been forced to stay out of the gym, her focus on her school work has intensified. She's a girl who likes to set goals, and if she can't do that physically, she will do it mentally. Right now, she is a sophomore straight A student. She takes all honors classes. Her plan after surgery next week is to concentrate on getting better physically. She does not hope to get back to cheerleading on a team until she is a senior in high school. She believes this will give her the time she needs to physically prepare with a private coach to get back onto a competitive team. For her junior year, any time spent not rehabilitating and working out will be spent on a difficult class load and community service. Her goal is to get the highest grade point average possible during her junior year, so that her senior year can be more fun without the stress of worrying where she can attend college. If she continues on the path that she is on, I fully expect her to attend the college of her choice, most likely on an academic scholarship.

A few months after Bri's PAO surgery, which went very well, Cynthia adds:

Bri said she won't be happy until she knows if she can cheer [do cheerleading] again. When I asked if maybe she'd like to tell people that the hardest part was the fear of the unknown before the surgery, she said no, this part has been equally hard. She's been feeling great for about six weeks and she still can't do the things she really wants to do. As an adult, I think I would be thrilled in the little things. As a teenager, Bri is just anxious to get on with it. We're really not allowed to ask her about her hip. She's not in pain and she feels that's all we need to know. I think there must be some anger about the time that has been taken from her. She's not ready to celebrate being pain free until she's in the gym again doing back flips.

As far as support goes, I was Bri's right hand man. She didn't receive a lot of support from her coach and teammates, and in fact, ended up changing schools before the surgery and changing gyms after the fact. That was emotionally tough. Her new school was very helpful — to a degree.

Individual teachers really made a huge difference. As a parent, I would tell people that the best thing they can do for their child is to make sure that home tutoring is set in place for when they are recovering. We are part of the American public school system. I knew, and several other parents have confirmed that it's the same in other states, that if my daughter was out for more than two weeks, the school would send us a tutor who could coordinate her make up work and keep her on track.

Because our doctor told us that she would be back in school at two weeks, I didn't set it up. Big mistake. She ended up being out a total of three weeks and only went back to school part time for about two weeks after that. It was very difficult to coordinate all the work with all the teachers. Some were very understanding, but others were very unsympathetic. Having one person at the school who was looking out for her would have been a tremendous help.

With all that being said, I'm happy to report that Bri ended the year just fine. When I called the school a month before the surgery to inform them of what would be happening, the first person I spoke with told me I should un-enroll her and put her in virtual school. She told me that if I didn't, in all likelihood, Bri would fail the semester. We refused to take this as a solution (just as we refused so many doctor's opinions before we found the right one!), and I'm happy to report that not only did she not fail, her grade point average (GPA) actually went up. She will start her junior year of high school with a 4.9 GPA, and has signed up for four Advanced Placement classes in the fall. Never underestimate the power of a determined individual!

Chapter 9
Other People

Unfortunately, our troublesome hips have an impact on our nearest and dearest, as well as the people we interact with at work and school, and society at large. In this chapter, we cover a little of what can go on in the family and further afield, and what sort of problems might beset you when you're dealing with your hip problems.

Genetic consequences

On a practical level, you may be feeling anxious and scared about what consequences your dysplasia may have on your children. Hip dysplasia does appear to have some genetic component. If you have it, your children are at a higher risk of developing it as well, especially your daughters. It is quite normal to feel anxious about passing this condition on to your children. However, while the risk is higher, it's still a *very* slight risk, around 16 in 1,000, as mentioned in Chapter 1. It's *most* likely that your children will be fine, although it's always wise to let their doctors know about your hip history, have them double-checked for hip problems when they are babies, and to keep an eye on them as they grow.

Responses to DDH

We'd all love to have wonderfully supportive family and friends, who are well-informed, who remember everything you tell them, and are there with tea and sympathy when you need it. You might be lucky, and be in this situation, too! However, the reality can be quite different. People forget what you've told them, don't have the emotional or physical energy or inclination to try to understand your situation, without necessarily meaning to be that way, and can be even downright rude and callous at times.

If you have some friends or family who really don't 'get it', and either belittle your fears and experiences, or want to keep on talking about it when it's clearly distressing you, you may need to get a bit of distance from them until after your surgery. Either try to explain clearly and with non-confrontational language ("I" statements such as "I am feeling upset", are better than "You" ones such as "You are being horrible to me!"), or just arrange your life so you don't need to interact with them quite so much.

If they really are unavoidable, then try your best to either explain the cause of your distress, or build up a psychological barrier, and don't let their words intrude. Easier said than done, we know, but when it comes down to it, they don't have a clue what they're talking about, and you can safely ignore them.

If you *do* find people around you who are supportive, understanding, and compassionate — hang onto them, and treat them well! You will find many great supportive and understanding communities online, too, we discuss them in more detail in Chapter 17.

Chronic pain and reactions to it

Acute pain is the sort of sharp, instant pain you get from an injury — a bee sting, cut finger, or a burn, for example. On the other hand, *chronic pain* is pain that lasts, and lasts, and keeps coming back. This is the sort of pain that us hip patients generally have to live with. The word "chronic" means "time" (from the Greek word *kronos*). When pain is constant, it affects everything about your life. It hurts to move, it hurts to be still, it hurts when you're trying to sleep. No position is comfortable. Your temper suffers, and this can impact on your relationships too.

Pain is a very subjective (personal) experience. The exact same injury or condition may be quite tolerable to one person, and agonising to another. There are many factors that determine how we cope with and experience pain, from our mood on a particular day, to stresses we're under at work, and other conditions we have, whether chronic (e.g. fibromyalgia) or acute (e.g. the flu). We discuss how to manage chronic pain further in Chapter 3.

Psychological treatment for chronic pain can be quite effective. Cognitive behaviour therapy, hypnosis, relaxation, biofeedback, and supportive counselling are some effective measures[62]. They work best when used in parallel with medication and/or physical therapy for pain.

With chronic pain, you may want or need to try to avoid relying on pain medication for lengthy periods of time. Some over the counter medications can be problematic if taken for long times. Others are quite safe for long term use. Ask your doctor for help, they're the expert in this sort of thing. Your local pharmacist will be helpful too.

One major problem arising from chronic pain is depression. If you're having difficulty making decisions, have lost interest in things you normally enjoy, are focusing on negative thoughts, have trouble managing daily life and chores, and feel like you're a useless burden, then it would be a good idea to see your GP about getting some help. There are some more ideas on dealing with depression in Chapter 7.

As if all this — chronic pain, disability, depression — wasn't enough, you may find yourself on the receiving end of some very odd reactions from acquaintances and others.

Unfortunately, the public at large can have some rather unhelpful responses to your pain and disability. Especially when the problem is internal and invisible, people can be surprisingly callous and unsympathetic. The underlying — and generally unconscious — thoughts they may be having run along the lines of "Is this person *really* sick, or making it up? I can't *see* anything wrong with them. Are they faking it and trying to trick me?" People doubt the validity of your condition, and go into "self-protection" mode, rather than offering help or sympathy[63].

You will probably find that the only people who *really* understand, are those who also live with chronic illness or pain of some sort. Clear visual cues (e.g. using a walking stick, or wearing a 'Hip Chick' t-shirt with some obvious slogan) can help to explain to people why you're limping, walking slowly, doing the "old lady" exercises at the pool, or using that disabled parking spot.

I've had several strangers comment, "You don't need those!", because I was able to go so fast on my Milliennial crutches.

My 7 year old daughter broke her ankle, and it was a sight to see the two of us crutching side-by-side. People thought we were joking. **Lara, 40 USA**

The general social response to disability and pain can be quite disheartening at times. By being disabled and in pain, you are unconsciously and unintentionally presenting a demand on others; you need special treatment, consideration, and adaptations, you're different, you need help. People often feel quite inadequate to meet that challenge, and don't know what to say or do.

One small thing in favour of hip dysplasia is that, as chronic conditions go, at least it's pretty socially acceptable, and can be a more obvious disability, especially if you're using a walking stick or crutches, or are wearing clothes where your scars are visible (like swim suits). People are more likely to believe you, and may be less likely to feel uncomfortable discussing it, unlike if you have endometriosis or HIV. But it's a small concession.

People stared. I just ignored them. People helped with opening doors, that was good. One girl cut in line in front of me at a bathroom at the beach. This was rude for two reasons: 1. She cut in line. 2. I was on crutches! **Shilpa, 27 USA**

If they say something rude, don't stoop down to their level and say something rude, too. Ignore them. You have to stay above them. **Bri, 15 USA**

Keep your cool! Remember no-one will truly understand what you are living with, so their hurtful comments or snide remarks are only down to ignorance! **Rhianna, 28 UK**

I have experience responding to comments or dealing with the effects of caring for other massive health issues (in a family member) and through discussions with others, gained the following wisdom: Others will *never* be able to truly understand what you're going through because they haven't lived through what you are living through.

148

It may truly be too much to expect, because **their** lives are not changed; they are not experiencing the chronic pain, nor having life-changing moments, or making moment-by-moment decisions like you are. It's not that they don't *want* to understand, they probably *can't* understand, because the words we use only convey a portion of the experience.

One of the results of this communication gap is that when a listener hears about your hip problem, and you give them an overview or brief explanation, they feel like they *do* understand the issue. Many listeners are familiar with the concept of an elderly person getting a hip replaced, so that's what they equate your issues to. In their mind, it's as uncomplicated as that, even though what they're hearing from you is that you're an anomaly, and they assume that's because you're too young to have worn out your hips. In light of the age discrepancy, they make what they feel is a reasonable adjustment in their understanding, and it's still fairly simple in their minds.

So, it can be a case of they *want* to understand, and then they think they *do* understand, when really they only heard or grasped a very small piece of the whole picture.

Once again, it's an issue of "They don't know that they don't know" all the details, specifics, or the resulting limitations, or effects of hip dysplasia on your life. They may even perceive that you're too young to have what they know is an elderly population problem, so they assume you're waiting "until it gets bad enough to do something about it". They may make the error of thinking that means that the pain or limitations are a non-issue, equating "little-to-no damage" with "little-to-no pain/limitation". Those who are experiencing hip pain or problems know that the two are not always closely correlated, as bodies and minds are amazing in their ability to keep going in spite of pain or problems, until enough is enough. **Brenda, 48 USA**

I get stupid comments from people in work all of the time. Some examples are: "Limping well today", "How's the gammy leg?", "Why on earth are you taking the lift, you should be walking". A lot of people in work have also commented on my weight increasing, because I am not able to exercise half as much as I used to. I try to brush them off, answer with whatever … and convince myself that if it were them in this situation, they would not be coping as well as I am (even though that might not be the case at all).

I just keep thinking that I could easily not be in work, but I am and I'm keeping going, which makes me feel a bit stronger. I also think that they obviously don't know the full impact of the condition and how much if affects me. I am sure that if they did, if I told them, that they wouldn't make such comments. I don't want to go around telling everyone though, so I just brush them off.

What does upset me is when my family or friends suggest to do something — or organise trips, hen's nights, and so on, which involve something that I will find difficult; I find that very upsetting. For example, a night out dancing. Standing still is very painful for me and so is dancing, so I would need a seat, which cannot be guaranteed when you have a night out! Also going for walks or staying at a friends — on the sofa — I just can't sleep on a sofa now, I get far too stiff and painful. I feel like an old woman and a bore saying I can't do that, I can't do this. When you explain people feel bad, but then understand. Sometimes it is hard having to say no and explain it all the time. **Sian, 29 UK**

Partners

With a supportive partner, your hip journey can be something that makes you stronger as a couple, and brings you closer together. Your partner may feel more needed than usual, and relish the chance to help you out more, especially if you're usually the organised one around home! But even the most loving and supportive partner can sometimes trip up.

My husband was always positive and sensible, but I did have to make him realise that once in a while, I did need to vent a bit, and just have him say "Oh dear, poor you, yes it is a horrible thing to have to go through". **Freja, 45 UK**

Unfortunately, for many reasons including lack of knowledge, or their own insecurities, past experiences, and prejudices, sometimes your partner's response can be much *less* than what you were hoping for. One young HipWoman's boyfriend told her he didn't want to be with a 'spastic' when he learnt of her diagnosis, and her boyfriend's father keeps telling her there's nothing wrong with her hips, and that she just needs to walk more, and that she can't fool the doctors forever!

150

Your partner may tell you not to worry too much until you've seen a specialist, and not want to discuss details of what might happen until it's absolutely final. This may arise from their desire for you not to get "too upset" about what is coming, until you're absolutely certain that is what is happening, but it can be really frustrating, and make you feel rather unsupported.

With my husband, I was met with a sort of "Until you see Mr J there is no point worrying about what might not happen" sort of attitude, which I didn't really find terribly helpful! I spent hours searching on the internet, researching, and reading. The more I read, the more I realised that I was either up for a PAO or a THR, but still I got the "It might not be that bad" attitude. I had seen the x-rays so I knew where I was heading with this, and I felt rather alone in my conclusion. I did find this upsetting, but then I understood that my husband didn't want me to have to go through anything like this. His way of dealing with it was to not acknowledge it until I had been referred to Mr J, if that's where it was all heading — until that point he didn't see the point in getting worked up about it. **Annick, 47 UK**

Your partner may simply misunderstand the seriousness of your condition, and that surgery really *is* the only treatment option in most cases. They may think you are overreacting and jumping to conclusions, when in fact you're following your doctor's advice. They see you every day, after all, and may think you're coping okay — especially if your hips aren't too bad yet and you're looking at a PAO. The leap from "You look okay to me" to "You need major hip surgery" is an understandably big — and often counterintuitive — one.

If this is the case, education is the solution; sit down with them, and explain the anatomy of your "special" hips. Try not to take their comments personally, and just state the facts of the matter. Prefacing sentences with "My surgeon says …" can help. Show them your x-rays. Diagrams and books (like this one!) can help too.

Another cause of odd reactions, and lack of support, can arise if your partner has their own bugbears to contend with, in relation to past experiences, or fear about what you have to go through. They may be afraid of losing you on the operating table, or have seen a close relative suffer in hospital. They may be feeling completely helpless to help you, and not know what to do. They may be feeling not up to the challenge of looking after you after surgery, in addition to everything else that's on their plate.

If they've had surgery themselves in the past, or traumatic experiences in hospitals or with the medical community, they will be dealing with all those past experiences, too, when it comes to handling your situation. They may overreact to unexpected things, like going to pre-admission clinic at the hospital. There may be times when you'd really like them to be there for you, and they just can't cope with it. Your increasing disability may be scaring them more than they care to say. Your decreasing sex life (if that's happening) can be another cause of conflict, frustration, and misunderstanding (we discuss sex and DDH in Chapter 16).

Again, this is a time to sit down in a quiet place and discuss the problem openly. Use "I" statements like "I feel unsupported at the moment …" rather than accusatory language like "You're letting me down, what's wrong with you?!" You may need to listen to their stories from the past, ask questions such as "Why do you think you're so anxious about my surgery?", and allow them to express their concerns. Some anxieties can be addressed with information Your partner may also like to read the chapter we've written for carers (Chapter 18).

Children

It can be hard on kids having a parent who is struggling with increasing disability and chronic pain. What you're able to do with them may be changing — fewer walks to the playground, perhaps — and that can be hard for them to understand. They might be worried deep down that *they've* done something to cause this.

Their friends at school might ask about "Why is your mum on crutches, did she break her leg?" or "Why does your dad walk like an old man with that stick?" These comments might worry or confuse them, they may not know what to say.

It's important for kids to not feel emotionally or socially isolated because of your pain, mobility, and hospitalisation. They still need to spend "fun time" with you and others — regardless of their age. Think of proactive ways you can still engage with your children, maybe help them learn a new craft or skill like cooking, read them stories, play games with them, talk with them.

YOUR HOSPITALISATION

While they may hide it, many children will have fears for your safety and suffer from separation anxiety when you go into hospital. It's important that they are kept in the loop with what's going on, this will help to ease their worries. How you explain your problems and treatment to your children will depend on their ages, you will know best what suits each child. We've included a few general guidelines below.

Positive and definite answers are always better responses for kids rather than vague or abstract answers to their questions. For example, "I will be fine!" (and you will!) is a better thing to say rather than "The surgeon thinks I should be okay after my hip replacement, as long as I don't get any complications". Give your children a rough time scale, so they know when you will be back home from hospital.

It's a good idea to arrange for your children to visit you in hospital, which will reassure them that you are fine — better than them worrying at nights and conjuring up all kinds of upsetting scenarios. If you can, arrange for a brief visit to the orthopaedic ward with them *before* you actually go in for surgery, so they have a good idea of where you will be.

Arrange things so that their regular routine at home can be maintained while you're away as much as possible, so that life at home is predictable.

Make sure they know you love them, reassure them. Focus on the benefits, tell them about how after you've recovered from the surgery you will be able to do so much more with them, like giving them baths, going for walks in the park, going to the playground, and so on. They can already see that you have trouble with mobility, and can't run around or play on the floor with them like other parents

If you have a range of ages in house, tell your eldest children first, and then more specific information for each child individually at appropriate times, with the shortest notice of your surgery given to the youngest.

THE LITTLIES

For very young children, an explanation as simple as "Mummy has a sore hip, and needs to have it fixed" may be plenty. If they ask questions, respond with simple but straightforward answers. It's important that if you're having surgery, they're aware that you're not being "punished" for anything, and that the surgery will make your hip better, even though it will be more sore for a while. Reassure your little ones that they can't "catch" bad hips from you (like a cold), and that it isn't *their* fault that your hips are bad or that you're in pain.

You may see regressive behaviour when they're stressed, so be on the look out for more clingy behaviour, fear of the dark, thumb sucking, and so on.

Pre-school children will only need a few days' notice that you're going into hospital, and very simple explanations.

My children were just turning 2 and 4 years old when I had my PAO. They already knew I had a sore hip, so I told them I was going to hospital so a doctor could make it better. The oldest asked if he would cut me and I said yes, but it would not hurt because I would have lots of medicine, but I would be sore for a while. I think honesty is the best policy. I also took them with me when I got admitted so they knew exactly where I would be. **Lea-Anne, 40 Australia**

I explained Mommy's hip bones were in the wrong place and hurting me. The doctors would make me sleep for two hours while they fixed them. Then I would be able to walk and stand without pain. **Lara, 40 USA**

I haven't told so much about my hips to my children because they are rather young to explain the whole process. But they've seen me walking, getting tired ten minutes after the beginning, with pain and stiffness in my movements. **Alvaro, 38, Spain**

Caring for very young children in the weeks after surgery will be a challenge for you. Initially you will probably need to have someone come in to help out, especially if you are a single parent. Otherwise, your partner will need to shoulder most of the child care duties for a while.

I think it is better if somebody else is spending time with your children when you're in hospital and recovering, and they don't feel abandoned. I am a single mom, and my mom was staying with me during recovery, spending time with me and my daughter equally. I think it is great to get kids involved in taking care of you, this way they feel like they contribute to you feeling better, and it means a lot! It did for my daughter! **Arpine, 28 USA**

During post-op recovery, it is so helpful to have someone at home to help take care of the kids. I made sure I was there to give them comfort and kisses when they needed me. Homework and reading books is something you can easily do with them while you are off your feet. I gave them "jobs" to do to help me, and paid them in cash or small prizes I'd bought beforehand.

Honestly, they coped a lot better than I expected. I worried so much about them prior to my surgery; who would bathe them, get them to dance class, find their missing shoes, and so on. My 2 ½ yr old still slept with us, and she would cry if I was not there by her side when she fell asleep each night ... I was more worried about them than myself! Thankfully, it turned out to be a lot harder on *me*, rather than *them*. **Lara, 40 USA**

OLDER CHILDREN AND TEENAGERS

Older children can have more notice of when you're going in to hospital, and greater detail about what's going to happen (but not too much!). They will still need reassurance that you'll be fine, and that the surgery will make things better for you in the long run. This extends further for teenagers, who can be fully informed about your schedule and surgery.

While they may put on a bit of a bluff, teens are likely to be anxious about your hospitalisation and surgery, even if they don't show it. You may see an increase in oppositional, emotional, or difficult behaviours, which actually reflect an increase in their worries. Take the time to sit down with them at a café, or some other "neutral" public territory where you can talk without things getting heated. Sometimes talking while being active doing something else (washing the dishes, working in the garden, driving somewhere) can work well. It's important that they know that you're listening to their worries, and taking them seriously. You might like to show them some of online information about the safety of anaesthetics (see Chapter 13), and other facts to counteract their fears.

Kids and teens can *sometimes* respond well if you ask them for help, and let them know that you will be needing to rely on their assistance around the home while you're in hospital and recovering. Make your requests for help very specific (e.g. "It would be a huge help if you could cook dinner once a week. What day would work best for you?") rather than demanding, nebulous, and somewhat overwhelming (e.g. "You'll have to do lots more around the house while I'm in hospital and on crutches!").

Written chore rosters can help here, both with clarifying your expectations, and dealing with a lack of support if your teen is non-compliant. A set time-frame (e.g. "We only need to do this for six weeks, or until I'm off my crutches.") can help as well, so these changes don't seem so permanent and threatening. Don't expect *too* much, though!

Areas they may be able to help in include :

◊ Fetching and carrying for you

◊ Bringing you drinks and snacks during the day

◊ Care of younger siblings

◊ Care of pets

◊ Making school lunches

◊ Cooking a simple meal

◊ Buying groceries (you provide the list and money)

◊ Emptying rubbish bins

◊ Tidying up in the living areas and kitchen

◊ Washing dishes

◊ Packing/unpacking the dishwasher

◊ Vacuuming

◊ Mowing the lawn and weeding

◊ Washing floors

◊ Cleaning the toilet and bathroom

◊ Washing clothes

◊ Hanging out laundry

◊ Folding and putting away clean laundry

A word about your children visiting you in hospital *after* surgery — wait for a day or two until you're over the worst of the effects of the surgery. You can be pretty sick in those first 24–48 hours, with vomiting from the anaesthetic, unable to move easily, IV lines in, machines that go "ping", catheters, and doped out on pain medications.

It's bad enough living through it yourself, but it can be quite scary for your children to witness, even for your "world-weary" teenagers. Best to recover a bit first, and then let them visit. It's a good idea for you or your partner to warn the kids that you'll be looking rather rough, and feeling pretty sick and dopey, with a tube in your arm, or whatever you have, when they first visit. It's easy for them to feel *more* worried or fearful, expecting the worst, if they feel you're hiding something from them.

It can also be a good idea to have some special new toys wrapped and ready to give to younger kids when they arrive at your hospital bedside, too, to keep them a bit distracted from your situation, and keep them quiet if you're sharing a room with other patients (who are likely to be elderly).

I have a daughter who was 6 at the time of my PAO. I didn't go into any details with her, just told her that my hip hurt and I needed surgery to get it fixed, and that she couldn't push me, and that she had to be careful with me and my hips because of that. **Arpine, 28 USA**

I was rather surprised to find out (over a year later) that my "cool as a cucumber" 18 year old daughter had found it "pretty distressing and confronting" to see me so pale and sick in hospital just after my THR. She said, "It's always really upsetting seeing a parent so bad." **Denise, 47 Australia**

In my experience children frequently understand far more than they are given credit for. Sometimes in the desire to protect children, adults keep information from them. I've found that this generally creates more fear and anxiety for children. Even if information must be adjusted for age appropriateness, I believe

157

it to be crucial to be as open and clear about it as possible. It inoculates them against unnecessary emotional trauma, which is especially important if this is something they must deal with for life!

In my personal experience as a child, one grandmother took me to the surgeon's appointment, where she was told I needed surgery right away. I was left out of the discussion and wanted to know what was happening. We joined my other grandmother to wait for my parents, and she had a different method of dealing with me. She matter-of-factly explained what was wrong, and that surgery would be needed. Though I wasn't excited, I understood what was wrong and what the plan was for it. That alone gave me a lot of comfort.

Now, 32 years later, I've recently had a THR, and my 5 year old niece and 7 year old nephew have been fascinated. They've checked out my staples and scar … and pronounced me cool! They understand the idea that the bones were worn out and needed to be replaced, and didn't bat an eye. My giant "boo boo" has been a bonding opportunity and a great learning experience too!
Katie, 47 USA

Talking to others

Once you've got a clear diagnosis, the time will come when you need to tell the other people in your life about what's going on with your hips. Some of these people will be your support network in the months ahead, so it's important to communicate clearly with them.

There are various ways to approach telling your friends, family, colleagues and others. Some HipWomen like to go for the "shock and gore" approach, giving full bloody details about impending surgery, in the hopes that shocking their audience will encourage them to remember what has been said!

It's inevitable — when you have a rare medical condition, you can't expect other people to be able to understand what you're going through, or for them to remember that you still have hip problems a month later. This is just human nature, and something that anyone with any chronic condition comes across. You are naturally the expert in your condition — and let's face it, you wouldn't be an expert if you *didn't* have hip dysplasia. So try not

158

to be too hard on your friends and family if they ask stupid questions, or forget that you have trouble walking far. You probably forget things about their sarcoidosis, epilepsy, or Crohn's disease, after all.

Many people like to try to educate their audience, and may spend hours explaining in careful detail about DDH anatomy, deformity, and treatment. This effort may be wasted, however, as people won't necessarily take it all in — but hopefully, some of the information will stick. Choose the more important people in your life for this approach — your parents, older kids, siblings, closest friends, favourite teacher, or boss.

The "fist in cupped hand" model of a hip socket is a tried and true method for explaining how a hip socket works. Then flatten your cupped "acetabulum" hand to demonstrate a deformed or shallow hip socket, graphically showing how the head of the femur (your fist) slips out of the socket as a result. You also might like to use the "egg cup and spoon" model from Chapter 1.

We have free PDFs of "explanation cards" about hip dysplasia, PAO, resurfacing, and THR you can download and print out, to give out to people, if you feel so inclined. < sutherland-studios.com.au >

With telling friends, I just told them my layman's version of "hip socket reconstruction", accompanied with the right hand clenched in a fist, cupped by the left hand to show how the socket was and how it would be after the op. Most of them didn't really believe me — I think they were in shock!
Annick, 47 UK

Some people also like to "explain away" your condition, through their own (generally faulty) assumptions and limited knowledge. This can be *extremely* annoying at times, being told "You just need to do more exercise" or "No wonder, all that ice skating/horse riding/dancing/hockey/child bearing has ruined your hips." Or that *they* know the best treatment for hip arthritis (ineffective copper bracelets, chiropractic treatments, or magnets, anyone?), or that their grandmother had a hip replaced and it was great and easy. This one can be *seriously* annoying! We bet their grandmother didn't have complicated hip deformities and weird anatomy!. Or they might say that "It really isn't that bad", and "What are you considering surgery for?" Aaaargh!

I got a lot of the, "My aunt just had her hip replaced too", "You're too young for that!" and, best of all, "It was having all those kids that messed up your hips, wasn't it?" **Lara, 40 USA**

When I initially told people about my hip dysplasia, many people attributed my hip pain to my years of ballet training and dancing in general. I understand that people are always trying to find ways to explain things to themselves, but I did feel like they were blaming me and my active lifestyle for my hip problems. I would respond to them by saying that I was born with hip dysplasia and that maybe ballet actually strengthened my hips instead of hurting them. **Melanie, 29 USA**

My experience has been that friends don't care too much about bones, people think that is not life threatening, so it is of no importance to comment on so much, so I've been a bit wary not to be boring. **Alvaro, 38 Spain**

Most people were shocked about my PAO surgery, and some compared it to a THR, which I was always quick to explain was not quite the same. **Lea-Anne, 40 Australia**

I told my family brief information initially, then I printed out information for them so they could understand it. At one point at work, when I first found out I needed a PAO, I found a documentary film which shows the operation. A colleague asked me what I was watching, so I did go into the full gore at that point — I think this was because I was in shock, it was the day after I found out I needed two PAOs. **Sian, 29 UK**

Most people, most of the time, forget what you've told them about your health. People forget you're in pain. You will have to repeatedly answer questions about why you're limping or using walking aids. You will have to explain your surgery a million times. Don't be surprised, we've all experienced this, it's just what happens, it's just the way people are. Our information cards (see our web site < sutherland-studios.com.au >) can help with this —carry some in your purse or wallet, and hand them out when being questioned for the umpteenth time.

If people knew I had hip issues, I just told them that I finally had a diagnosis of hip dysplasia, and that I would need a PAO (the full name "periacetabular osteotomy" always got a "What?!"), which involved cutting the hip in three places, rotating it, and screwing it back together. **Lea-Anne, 40 Australia**

Initially I felt that if I explained in great detail the negative effects that hip problems imposed on my life, and then similarly explained all the various attempts to get an accurate diagnosis, and then follow-up with the gory details of what was involved with solving the issues (PAO or THR), it would resonate with the listener. More often than not, it didn't resonate whatsoever, but rather bored the listener, or I lost their interest along the way. Now I generally say, "I have a crappy hip" to explain my limp and I say, "I'm waiting on a good time to have surgery to fix it", if someone expresses any interest. **Brenda, 48 USA**

I found that the best approach to dealing with ignorant comments is to educate people, and to clearly explain the diagnosis and surgery. If people were insensitive or forgetful, I would go for the full gore, and remind them that I had bones broken on a weight-bearing joint, and that I have enormous screws in my hip holding those bones together. I also posted my post-PAO x-rays on Facebook, so that my friends, family, and co-workers could actually see what was done to my hip. However, overall, my friends, family, and co-workers were extremely helpful and thoughtful, perhaps because I made sure that they knew the severity of my surgery. **Melanie, 29 USA**

I didn't like the fact that I had to talk about it over and over again, yet understood that I would have asked others as well. Some people excused their curiosity, some were pretty blatant. I found that most questions were driven by curiosity rather than desire to check on my well being. Human nature :)

I am the kind who takes everything very personally and gets upset easily, yet never talks about it or comes back with anything. I generally settled with the fact that most people don't really care about my hip problems, and tried to focus my emotions around those who did. **Arpine, 28 USA**

I didn't really bother responding to anyone who came out with something crass, because I felt they just didn't fully understand what was wrong and what I was going to have to go through to correct it. **Annick, 47 UK**

Most people I told were aware of my limp or the sudden weakness I might experience upon rising from a chair, so they were generally interested in hearing that there was a medical explanation that was being addressed. The comment that surprised me the most came from a nurse, who is also the mother of a young man who had a hip replaced in his 20's due to bone death from **avascular necrosis** (AVN): "I thought only babies had dysplasia." Her tone was somewhat condescending, and the look on her face implied that my explanation was not truthful or believable. **Brenda, 48 USA**

Based on people's reactions, this has become my official explanation of my hip dysplasia and PAO: "I was born with my hip joint sticking partially out of the socket. I didn't know about it until my late 20's when the uneven coverage of my hip joint started to wear down the cartilage and my hip started to hurt. To fix the problem, I had to have three of the bones in my pelvis broken, reshaped around the hip joint, and screwed back together. When the bones grow back together, my hip will be fixed and even stronger than an artificial hip."

I guess this would be the somewhat 'gory' response, but I ended up explaining it to people this way so that they would know that I had to go through major surgery with a long recovery time. Especially my co-workers, who expected recovery time for "hip surgery" to be shorter. My response also addresses that fact that I was born with hip dysplasia, so my hip problems are not the fault of my ballet training. And it also addresses why PAO was a better option for me than THR. **Melanie, 29 USA**

I didn't really explain much about it. I just said that I had a problem with my hip and needed surgery for it. When asked why and what kinds of problems, I said that the bones weren't positioned properly, so they had to be cut and re-positioned (I had a PAO). That usually did it. **Arpine, 28 USA**

Here are some more explanations you might like to use:

◊ I was born without proper hip joints.

◊ I had a dislocated hip when I was born and it wore out really fast, so I needed a replacement.

◊ My pelvis was reconstructed to give me new hip joints, as I am too young for a replacement.

◊ I've had major surgery on my hips, the recovery period is very long and I've had some complications and set-backs.

◊ I've had my pelvis broken 'x' times to give me new hip joints.

◊ I have arthritis in my hip +/- because I have a condition called developmental dysplasia of the hips.

◊ My hip sockets were too small and in the wrong position, so they have to cut them out of my pelvis and reconstruct them in the right place, with screws to hold my bones back together.

Workplace negotiations

Explaining your situation at work can be tricky, especially if you need a lot of time off work, and special considerations or modifications to your workload.

Unlike some "invisible" chronic conditions, you don't generally have much of a choice about whether or not you tell your boss about your hip problems — it's generally all too obvious!

It is best if your immediate supervisor and colleagues know about your DDH. You need to communicate clearly with your boss, especially, about the ramificiations of your increasing disability, any upcoming surgery, and your expected recovery times.

Part of this comes down to how well you understand and have accepted your condition. By reconciling yourself to your situation, and the limitations it may pose, you can present a fair and realistic case to your boss as to the restrictions and adaptations you may need in the workplace. Setting up chairs and tables for a conference is a job you probably need to avoid, for example.

Some things you may want to consider are whether you need any temporary adaptations to your workload, the ergonomics of your workplace, or your work hours. Using teleconferencing could be a better option for you rather than driving for hours to get to a meeting. Are there more supportive chairs that would make sitting for long periods less painful? Can you do some of your work from home?

If you find yourself getting more fatigued than usual, because of increasing hip pain, you may need to curtail overtime and going out at nights, so that you can still cope with the demands of your job. Say no to unreasonable demands, if you know that you really can't manage something at the moment.

Pacing yourself over the day can help — have mini-rests, and adapt your working situation so you can get through the day in better shape. Are there tasks you can do sitting down, rather than standing? See Chapter 7 for how to practice relaxation, meditation, and mindfulness — even a five minute "session" at your desk can help your mood and stress levels, which can help you cope with the pain.

One side effect of telling your employer and workmates about your hip problems may be that you suddenly receive more concern and solicitude. You may be treated differently from the rest of the staff. All of this may be welcome, of course, but some people feel awkward and embarrassed by this attention. If you work in an area which has a human resources department, it is important that you communicate clearly with them as well. You may find it helpful to have some letters from your GP, surgeon, and /or physiotherapist explaining your situation, and the adaptations that you need at work.

Basically, the more self-accepting and at ease you can be about your condition, the more at ease others will be around you. Your colleagues take cues from you about when you need help, and what adjustments to make, and when it's okay to discuss it[63].

If people don't understand why you're doing less, they may think you're not pulling your weight around the workplace, and get resentful. So it is important to take the time to communicate clearly. This creates more trust within the workplace in general, builds more support for you, and more understanding from them of your situation. But it isn't always an easy situation, especially if your hip problems are ongoing (as they generally are).

I'm a medical social worker, and I work closely with a liver transplant team. I experienced significant thigh and back pain when my hip joint was failing me. In addition, my limp was more noticeable and due to increased pain it impacted my sleep, daily activities, and work. Post hip replacement, I returned to work with reasonable accommodations, such as my work schedule was modified. Due to my physically and emotional demanding job, I reduced the amount of rounding with the doctors, standing and walking. My boss has been very supportive. But I do feel inadequate and guilty because I struggle to meet my job duties, and sometimes I need additional assistance with my work.

I had an ergonomics specialist come out to adjust my work station. I have a high chair with a seat cushion, back support, and foot rest. It helps, but I still need to get up every 30–45 min to stretch and walk around, due to stiffness and pain. My advice is to take all the time you need to recover before returning to work. I was eager to return back to my normal life and went back to work prematurely, which set back my recovery period. Also, transition slowly back to work. It helped me working part time for the first 6–8 weeks. I was exhausted just working four hours a day! Be patient and listen to your body. **Cathy, 39 USA**

At my work, my staff treated me right, understanding my situation and my dependence on crutches. Never crowding or sneering at me. **Alvaro, 38 Spain**

I felt lucky at work, as they were very supportive. I work as a senior engineer in a male dominated environment, so have never liked to draw attention or appear weak. My colleagues got used to me in meetings having to sit/stand/sit because of discomfort, or the resounding double crack on occasion (both hips!) when I stood up that everyone in the room could hear!

I do have to do some business travel and found that difficult, but my colleagues would help and carry my bags (whilst assuring me they weren't being patronising or sexist — which of course I knew, love 'em). The warehouse boys wouldn't — and still won't — let me carry anything, and as I recovered I got regular progress reports from the shop floor to senior management on how well I was walking — as if my physio had spies! **Dawn, 44 UK**

I am a registered nurse. Well for me, the right hip was always the "bad" hip pain wise, and although everyone at work was sympathetic enough, they just didn't understand when I kept having to get it fixed.

Surgery #3 on my right hip resulted in me losing my position because of policy, but I was allowed to keep my seniority for one year. Now I'm back in a similar position, with the possibility of yet more work on my right hip. Everyone seems to get it more now. Although I technically can't be on light duty (workplace policy), my co-workers do try to give me the lightest patient assignments. It's nice having that understanding, but I still feel bad sometimes for not always pulling my weight.

There is a possibility I'll need to find more "desk based" work versus floor work. There has been some frustration in this aspect for both myself and my husband, since my work doesn't recognize the need to place me in a less demanding job. It is kind of a dog eat dog world in that regard. If I'm well enough to be a nurse, I'm well enough to do any nurse job. Not exactly true, but it is what it is! Basically it boils down to seniority, and not taking care of your own. Staying on disability was not an option either, since it only pays 60% of wages after taxes, and you're only allowed 18 months total. It was just very frustrating to try to get everything in order sooner than I felt ready. **Marcie, 30 USA**

People at work were very considerate. When I was on two crutches, they would bring me coffee to my desk because I couldn't carry it. Would drive me to the gas station, and help me have my tank filled. Yet they didn't express pity, and didn't have that "You poor thing" attitude towards me, which I highly appreciated! Some co-workers who I don't work with directly and don't see quite often, used the hip as a conversation starter. "So how's the hip? How long are you going to be on crutches? Are you doing physical therapy? Does it hurt?" Same stuff, every day, over and over! Even now, seven months after my surgery, I get the "Howsthehip?" question! **Arpine, 28 USA**

I have been very lucky with my job and my direct manager. I work in a university, and they have been very supportive. The most important thing for me was when I returned to work, I did it on a phased return. I did four hours per day the first two weeks, then five hours each day for the next week, six the next, and so on. I had previously ordered an orthopeadic chair which is supportive. I also worked at my desk for the first two weeks, then built up going out on site, which I have to do for my job. They were also flexible with physio sessions and when I needed to see my consultant. I would recommend to anyone going to see Occupational Health/HR if they have any problems like this.

A girl I know, who had the same op as me on the same day, did have trouble going back to work. Her employer wouldn't make any adjustments for her, wouldn't do a risk assessment, or even give her a foot stool. Luckily and good on her, she stood up to them and said that they were not providing reasonable adjustments for her.

Part of my role in the uni is improving access for disabled staff and students, so I am very familiar with the relevant UK legislations. People with this hip condition should be treated by their employer as a disabled member of staff. As such, they are covered under the Equality Act 2010, and the employer is legally required to make reasonable adjustments for them. I would recommend if someone is having a lot of problems with their employer, to quote the Equality Act and remind the employers what their responsibilities are, and take it further if they do not. Usually, I find just that threat, or even just informing them you know what your rights are, is enough to get things done.

The thing I found difficult was people I didn't know all that well at work asking: "Oh, what have you done to yourself then?" When I would reply: "I've had hip surgery", they would generally say: "Aaah, you're young for a hip replacement". (I had a PAO.) It's nice that people care enough to ask, but some people make a joke about it, which is hard. I suppose it's just a reaction.

Having said all those positives about my work, I *did* feel very pressurised to go back to work. I was absolutely shattered starting back, and sitting still for long periods was very painful. I would recommend getting up and moving regularly if you have to sit for long periods at work.

I also didn't like people constantly watching me and the way I was walking, having a running commentry every day from other people about the way I'm walking — given I'm the one doing the walking, I don't need to be told how I'm doing! Comments about my walking stick I used also bothered me. In the most part I shrug it off, but sometimes it upsets me. **Sian, 29 UK**

My human resources department initially gave me trouble about not coming back to work sooner. I'm on my feet most of the day at work, and they actually said things like, "You're young, you'll heal faster than most people." and "Why can't you come back sooner and work on crutches?" I dealt with this in the following ways:

1. I had my doctor write me a note to stay out of work longer than expected. We both thought 3–4 months out of work was appropriate for my job, so he wrote me a note for 4 months. As my doctor said, "If you go back earlier, you'll be a hero." It's much easier to go back sooner than expected, instead of later than expected.

2. When I talked to the human resources department, I started every sentence with "My doctor says …", because the last thing they want to do is argue with a doctor. And I corresponded with them via email, so I would have everything in writing. Also, I was lucky enough to have a doctor who would put anything in writing regarding work restrictions to back me up, if I needed it.

3. I made sure that human resources knew the potential problems if I came back to work before I was physically ready. I said things like, "If I run around too much and get exhausted, I might fall and break my hip." Believe me, the last thing that a human resources department wants is for someone to get injured at work, so comments like this helped them understand the need for me to properly heal before I started working again. **Melanie, 29 USA**

We hope that on your hip journey you are surrounded by supportive friends, colleagues, and family, and that the suggestions in this chapter are helpful if any problems arise.

Chapter 10
Proactive Strategies

In this chapter we look at ways you can be proactive and involved in your hip journey. We cover emotional preparation for surgery, using information, keeping a medical journal, how to find a great surgeon, that nerve-wracking first visit, what questions to ask, getting second opinions, changing doctors, and how to take charge of your medical team.

Emotional preparation

Preparing for surgery takes more than just packing your suitcase and going to pre-admission clinic. You may be one of the lucky ones who isn't particularly bothered by the thought of surgery, but the vast majority of people have some level of fear and anxiety about the whole process.

Let's face it — we don't like having surgery! Who would? Knowing that you have surgery in your future – whether it's a year off, or next week — can be very hard to cope with. It's always there in the back of your mind, a little knot of worry and fear. Even though you know that the vast majority of people get through surgery perfectly fine, and with no complications, well, you might be different! It could all go wrong, couldn't it?! You could be that unlucky one in thousands … and so the knot of fear and anxiety stays. This is normal — most of us go through this level of emotional upset before our surgeries.

For some, however, these feelings can escalate. You might start having nightmares about surgery, feel driven to read horror hospital stories on the internet (don't do it!), or worse still, watch YouTube videos of surgery, which will only end in tears. Talking about your surgery makes you cry or panic. You start to prepare for not coming home from hospital, finishing off projects at work, making sure your will is up to date (well, that doesn't hurt, but it shouldn't be in reaction to morbid thinking). At this point, it really is time to step back, look at yourself (kindly and with compassion), and think about getting some professional help. Counsellors really can help, it's worth making that first difficult phone call for an appointment. Remember that you are able to enquire or "interview" a prospective counsellor or psychologist before you commit to seeing them regularly. Some good questions to ask yourself about them include : Do they have experience with PTSD/chronic pain patients? What style of therapy do they use (e.g. Cognitive Behaviour Therapy, Mindfulness), and am I comfortable with it?

During surgery, we have to put our very life in the hands of people who we don't know all that well. You've met your surgeon a few times, and the anaesthetist maybe once or twice? These people — and their teams — are going to be keeping you alive while you're vulnerable and unconscious, and doing incredibly intrusive and rather violent things to you while you're knocked out. "They'd better be bloody good at their jobs!" you think. It's hardly surprising you feel anxious.

We really have to trust that the system works, that your medical professionals are just that — professional, thoroughly trained, and knowing what they're doing. And really, the answer to all those doubts is **yes**, your surgeon and anaesthetist *are* incredibly clever and thoroughly well-trained. The rigors of medical school followed by years and years of specialist training in their chosen field, including tough examinations all the way through, means you really are in incredibly safe and skilled hands.

Chapter 7 has a lot of practical coping strategies for coping with anxiety and fear about surgery; in particular, you may like to check out the sections on meditation, mindfulness, guided imagery, and surgery scripts.

MEDICATION

If you are really struggling to get on top of severe anxiety and fear in the weeks leading up to your surgery, don't hesitate to talk to your GP about it. They can prescribe an anti-anxiety medication like Valium (diazepam) for you — if it works for you, it can be utterly wonderful and just make all that worry and fear disappear. Taken for such a short time it isn't addictive, and it can make life so much more pleasant for these hard times. Don't feel that you're being weak in getting this sort of "chemical assistance". You're having to be quite strong and brave enough fronting up for the surgery at all, don't feel you have to suffer through the fear as well.

Having Valium or similar on the day of surgery can also help a lot, ask your surgeon about this well in advance of your surgery, so they can make a note on your chart. It's vital that the anaesthetist knows what medication you've had before you go in to the operating theatre.

I was really struggling with my PTSD in the weeks leading up to surgery. I had a week on diazepam before surgery, and temazepam on the day, and it made such a vast difference, all my fear and anxiety just *went away!* Magic!
Denise, 47 Australia

Being informed

Information can be your saviour. Being well-informed, from reputable medical web sites, good books, and your medical team, can really help a great deal in quelling your fears.

For example, if you're scared about going under general anaesthetic, ask questions of your anaesthetist; jot them down in your medical notebook before your appointment. No question is "stupid", honestly, they're only too happy to help you out. You'll have an expert in anaesthetics right in front of you, after all, take advantage of the opportunity!

The excellent site All About Anaesthesia < allaboutanaesthesia.com.au > has a great in-depth section under 'About Anaesthesia', called 'Common Fears'. It's definitely worth checking out, if you have any anxiety or questions about going under general or regional anaesthetic.

The same goes for fears about surgery, your particular operation, and the risks of complications. Mind you, for some people knowing *exactly* what they're going to do to your hip can be a bit too much information. We've even seen the *partners* of hip patients get faint and need to lie down after hearing what a THR or PAO involves!

Sometimes glossing over this information is fine — be wary of watching medical TV shows that show these procedures, and especially avoid watching videos of the surgeries on YouTube. Even illustrations, photos, and diagrams can be a bit too much information for some. Just so long as you know enough to give your signature on the surgery consent form, you don't have to know more, or delve into the details of the procedure any further. This is an instance where we'd certainly advocate a bit of 'ignorance is bliss' if fear of the surgery is getting the better of you.

As a surgical trainee working in orthopaedics, I actually had to assist my consultant do a PAO on another patient, about a month before I had mine. Now, I want to be an orthopaedic surgeon, *love* operating, and normally have no problems with surgery — I am completely fascinated by it! However, this was one of the worst experiences in my career; there comes a point where too much knowledge and information is definitely a negative thing! However, having now had both my PAOs, I find myself very keen to see the procedure done again. **Sophie, 29 UK**

Keeping a medical journal

One very proactive thing you can do throughout all your hip travels and travails is to keep a medical journal. Just get yourself a small book with blank lined pages; hardcover is probably best for durability. Bring it with you to all medical appointments, and take notes either during or immediately after your appointment, while all the details are fresh in your mind.

This notebook is the place to write down your doctor's and surgeon's advice, make notes about medications you're on, record blood and other test results, write down questions to ask as you think of them, track symptoms (such as keeping a pain diary, see Chapter 3), note down the exercises your physiotherapist wants you to do, and so on. You might like to glue the business cards from your specialists and health professionals inside the front cover, so their contact details are always to hand.

These books are a valuable record both at the height of your hip treatment — your doctors will love that you'll be able to pinpoint the exact date you started on a certain medication, for example — and as a historical document, so you can go back and accurately find out what you did several years before, if needs be.

Your medical team

When you have any chronic, complicated, or rare medical condition — such as hip dysplasia — it's important to gather a supportive medical team. As the patient, you're the head of the team. You're the boss. You're responsible for your own health, with a great team of specialists working for you. Whether it's through your taxes or directly, you are paying them, after all!

On your hip team you may need some or all of:

◊ a general practitioner/primary care physician.

◊ an orthopaedic surgeon who is experienced with hip dysplasia.

◊ a sports physician (they are specialists who are skilled at helping with problems with bones, muscles, ligaments, and tendons).

◊ a physiotherapist/physical therapist.

◊ a podiatrist.

◊ an occupational therapist.

◊ an anaesthetist.

Never forget that you always have the power to "hire and fire". If, for example, you're really not happy with your physiotherapist, don't keep seeing them. There are almost always other options and alternatives, even if it means travelling further, or paying a bit more.

Some HipWomen have travelled interstate, or even to other countries in extreme cases, to get to the best hip surgeons. If your family doctor thinks you just have to "grin and bear the pain" from your hip, and isn't willing to refer you to a specialist — see a different doctor. If you don't feel confident in your local orthopaedic surgeon, or they belittle or upset you, you can choose to go further afield. If your physiotherapist isn't experienced with hip dysplasia, and is giving you exercises that your surgeon hasn't approved, change physios. Although, sometimes health insurance requirements can mean that you don't have much of a choice :

One surgeon constantly makes me feel like I'm a big, overly anxious stress basket because I did research on PAO, got second opinions, and asked questions. He told me my HipWomen support group was not helpful because

173

all it is "just a group of women complaining because they didn't get good outcomes" (not true!), and I was not getting accurate info. He never doubted I had anything wrong with my hips, however. I tried not to take him personally, as we observed he was also disrespectful to others, like the nurses and other staff. I would have changed docs but unfortunately, he is my link through my health insurance fund Kaiser to Dr. K, who did my PAOs. I still need to see him to keep my connection to Dr. K. **Lara, 40 USA**

Many orthopaedic surgeons, while they know about hip arthritis and hip replacement, haven't got a clue about dysplasia — many hippies have been told such things as "Your bones look normal", "It's a trapped nerve", "Just put up with the pain until you can't walk any more'", and "Here, have some cortisone, that'll fix things."

One of my first doctors told me ibuprofen and rest would do the trick, but from the pain I'd been experiencing, I knew there had to be something more. **Bri, 15 USA**

When I originally started experiencing pain in my hip, I was told by my GP that it was trochanteric bursitis. I had x-rays done, which got lost! Finally I had more x-rays done, got an appointment to see a rheumatologist, which was then changed to an appointment with an orthopaedic consultant at the hospital where I work.

He wouldn't even consider any kind of treatment because of my age: "I'm not prepared to take the risk." were his exact words. He referred me to another consultant in a town over three hours' journey away! I wasn't prepared to travel that far only to be told again that because of my age, there was nothing that could be done.

 Eventually I asked for a new referral from my GP, which was the best thing I could have done. I finally found a surgeon who specialised in the replacement of hip joints in younger patients. **Tina B, 36 UK**

I saw an orthopaedic surgeon, when I started to develop arthritis in my "bad" hip, in my 30s. He told me that nothing would help apart from a THR, and just to wait until the hip joint "collapsed", and then have THR surgery. He didn't even think I should bother with physiotherapy, and just told me to lose weight. When the time came for me to have my THR (over twelve years later), I chose a different surgeon! **Denise, 47 Australia**

I was disappointed that I wasted a year in seeing chiropractors who misread my x-rays and MRIs — and didn't give me the correct diagnosis. I felt upset that I still had to pay for their services! However, I was impressed by the first OS I met, as he was honest, and didn't want to take a chance to repair the labrum himself. He referred me to the group who I am now seeing. **Christelle, 38 USA**

We know this may not be easy, it may involve more expense, more travel, inconvenience, decision, and delays — but it's *your* health and *your* hip after all. It can be very empowering to suddenly realise you *don't* have to keep seeing that dropkick surgeon with no people skills who makes you feel two inches tall — you can go elsewhere! There are wonderful surgeons out there who are skilled, knowledgeable, *and* caring human beings to boot (and never underestimate the power of the Cute Factor!). Don't give up.

If you're not happy then be persistent! Keep pestering your doctor/consultant until you get some measure of satisfaction! **Tina B, 36 UK**

Keep looking, get as informed as you can, and believe in yourself because no-one knows your body as well as you do. If you can, take someone with you to medical appointments who will ask the tricky questions and stand up for you. **Lea-Anne, 40 Australia**

Do your research, hold your ground! No one else will fight for you better than yourself. Use the HipWomen forum to get advice/tips. It was *invaluable* to me. **Lara, 40 USA**

Go to someone else. You should trust the person who could potentially operate on you. If you don't have total and complete trust in them, there is probably a better doctor out there for you. **Bri, 15 USA**

Finding a good surgeon

Unfortunately, getting a clear diagnosis with these sorts of hip problems is no easy feat. You may have to see several doctors before finding one who can tell you what the problem is. This is a common situation for many hip dysplasia patients. DDH is a rare condition, and there are a limited number of places and surgeons that have the necessary experience to diagnose and treat DDH in adults. So, if your diagnosis is made by a non-specialist, it is important to be referred to one who specialises in dysplasia. If you were diagnosed and treated at birth, and are starting to have problems in adult life, make sure you are referred to someone with the necessary experience.

When I was diagnosed, I was in my third year post-qualification from medical school, and had worked in several different orthopaedic departments. However, my comment when I was told I had dysplasia was: "As in the condition babies have?"

Despite six years of medical school, and three years of further experience, I had never seen (or heard of for that matter) an adult with hip dysplasia. My point is that not all doctors will recognise hip dysplasia, and there are even fewer who have the expertise to treat it well. **Sophie, 29 UK**

LOCATING A SURGEON

Family doctors can refer you on to a specialist. In the UK the NHS has a system called "Choose and Book" which in theory enables patients to choose which specialist they would like to be referred to. It is important to be aware that specialists with experience in hip dysplasia are few and far between. This means that the best surgeon may be "out of area", which can make the referral process more complicated, but by no means impossible. This is also true for those in the USA, and insurance companies, where the specialist may be out of network. In Australia it is possible to nominate the surgeon you prefer within the Medicare system.

I had the choice of several orthopaedic surgeons for my THR; most of them were older men, with a great deal of experience, but they weren't dysplasia experts. In the end I chose a younger surgeon who had fewer years in practice, but he was an expert in dysplasia, and really understood my weird anatomy and childhood surgeries. He did a fantastic job. **Denise, 47 Australia**

Many orthopaedic surgeons will know a colleague that they can refer you on to. The internet is usually a good place to start. There are many support groups and several Facebook groups, where patients discuss their condition, surgeries, and which doctors they've been to. We also cover this topic in further detail in Chapter 17.

The HipWomen group on Yahoo, for example, is a mine of information, as other patients are happy to help you out, including tips on which surgeons they recommend, in a wide range of countries. Most of us don't know other hippies in real life; it's only through the internet that we've been able to connect with each other.

Look up medical journals related to diagnosis and treatment, and see who the authors are or where they are from. PubMed <www.ncbi.nlm.nih.gov/pubmed/>, which also encompasses literature from Medline, is a good place to start, and it's free. Once you have found some articles, a good tip is to search through the references of these articles to find even more articles. This can save you time searching. Here are our recommendations for things to try :

◊ Research medical papers written by surgeons who have an interest in hip dysplasia.

◊ HipWomen, Hip Chicks, and other online forums are a great resource, and can help you identify surgeons in your area.

◊ General orthopaedic surgeons, if they have the ability to diagnose hip dysplasia but not to treat it, can often recommend another surgeon you can see.

◊ The International Hip Dysplasia Institute in Florida, USA can recommend surgeons for you.

◊ The Pediatric Orthopedic Society of North America (POSNA) can advise on **paediatric** surgeons in your area for those living in the USA, if your child needs surgery.

177

For those in the UK and Australia, you may find you have to go through your GP to get a referral who can then recommend a specialist to see. You may need to see a general orthopaedic consultant first, as specialists in hip dysplasia sometimes require a referral from another specialist instead of from a GP.

Your first visit

When you see a surgeon for the first time, it's a good idea to have your medical journal with you (see pg 172), a good list of questions to ask (trust us, you'll forget some of them if you don't write them down beforehand!) and a support person — your partner, parent, or a close friend. They can take notes for you, offer moral support, ask things you might have forgotten, and just generally help you feel not so alone and anxious. Remember, you are interviewing the doctor, in effect, so if you don't feel comfortable with them, you don't have to make a repeat appointment.

Here's a list of possible questions you might like to ask your orthopaedic surgeon the first time you see him or her. This list is also available on our web site < sutherland-studios.com.au > as a free PDF download:

◊ Can you explain why my hips hurt?

◊ If I have dysplasia, how severe is it?

◊ Is it unilateral (one side only), or bilateral (both sides)?

◊ If it is bilateral, is one hip worse than the other?

◊ How can I manage my hip pain short-term and long-term?

◊ Are there non-medical things I can do in my life to help keep my hips healthy—what kind of exercise is good for me, is there anything I should *not* do?

◊ Do I have any labral tears? If so, please explain what that means and how they should be treated.

◊ Would you repair a labral tear/impingement?

◊ Do I have osteoarthritis? If so, how severe is it?

◊ Will my hip dysplasia be a problem if I get pregnant?

◊ What treatment do you recommend for my hips?

◊ If surgery is not recommended:

- Will I need surgery in the future?
- How many years in the future?
- How often should I come in to be checked?

◊ If surgery is recommended, please explain these types of surgery to me:

- Pelvic osteotomy (PAO)
- Total hip replacement (THR), and the different types of prostheses available
- Hip resurfacing

◊ How will the surgery that you recommend improve my hips?

◊ If I don't have a PAO now, will a THR or resurfacing work for me later on?

◊ If I have a PAO, do you remove the screws, and if so, when? How do you carry out this procudure?

◊ Am I a candidate for a resurfacing?

◊ If recommending a THR, which sort of prosthesis would you recommend and why?

◊ What approach do you use for a THR — posterior or anterior? What are the pros and cons for each approach?

◊ If I need both hips done, how long apart should have the surgeries?

◊ Can leg length differences be corrected, will there be a difference post-op?

◊ What are the best case and worst case results of the surgery?

◊ What goes "good outcome" mean to you?

◊ Can you give me rough recovery time frames for activities such as weight-bearing, getting pregnant, lifting children, driving, working?

◊ Can you supply an exercise plan to take to my physiotherapist?

◊ Will my health insurance cover this treatment?

Second opinions

It is your right to have as many opinions from as many specialists as you like, if you so desire. If you feel unhappy with your surgeon, for whatever reason, you are entitled to ask to be referred for a second opinion, or seek one out on your own. You are also not obliged to give your reasons. With such a major diagnosis, it is important to feel absolutely secure in your medical team.

Having such a complicated and unusual condition can mean that you may come across doctors, physiotherapists, sport physicians, and other health professionals who can be dismissive, abrupt, rude, or just plain wrong. Always remember that, while you do need to rely on a team of people to help you, *you* are the head of the team. You can "hire and fire". Find the best people possible, it really is worth the effort.

There are many reasons people seek a further opinion — here are some examples from DDH patients :

I got a second opinion because on my first visit to a surgeon I was told to suffer until the pain was unbearable and just in that case undergo a total hip replacement. **Alvaro, 38 Spain**

I was initially diagnosed by a sport medicine specialist, who then referred me on to a paediatric orthopaedic surgeon, who — once it was established I needed surgery — then referred me on to a surgeon who specialised in adult hip dysplasia. **Sophie, 29 UK**

I saw three orthopedic surgeons who all perform PAOs, regarding whether or not I was a good candidate for surgery. I wanted opinions from all three of them, because a PAO is a major surgery, and I wanted to make sure that I was making an informed decision.

Interestingly, all three surgeons gave me some small piece of information about my diagnosis or the surgery that the other two surgeons had left out. I feel lucky that all of these surgeons were less than an hour from where I live, so I was able see them easily. However, one of the surgeons was not covered by my insurance, so I had to pay for the visit out of pocket. I believe it was $300, but it was still worth it to me. **Melanie, 29 USA**

I had to seek a third opinion about my diagnosis, because the first two doctors could not figure out what my true problem was, and therefore I needed some answers that I was not getting. **Serena, 39 USA**

I sought out *four* doctors' opinions because I could not believe I needed my hips replaced, especially because at that time I thought I was way too young. I finally started believing it when I had a well-known physical therapist (who had *no* attachment to surgery) tell me, after seeing my x-rays, that the diagnosis and need for surgery were obvious, and "The fat lady had sung". **Jodi, 56 USA**

I continued to seek opinions regarding what was wrong with my hip, because every opinion seemed to be a series of guesses (hip strain, hip flexor weakness, piriformus inflammation, bursitis, sciatica, and so on). The recommended treatments, such as rest, anti-inflammatory meds, and physical therapy lead to very short-term results. **Brenda, 48 USA**

I had originally been diagnosed with fluid on the psoas tendon. For this I had had three cortisone injections under general anaesthetic, but they hadn't seemed to cure the problem. I was then advised by my first OS, Mr F, to undergo a 'psoas tendon release'. I had wanted to get a second opinion but didn't know how to go about asking my OS if it would be okay, because of fear of insulting him. I had spent time with my husband trying to come up with ways of saying it, but didn't actually find a solution! Luckily, when I had my next appointment, I asked him to go through the operation. As soon as he got to the "four inch incision in the groin", I asked him about keyhole. He, being as old as the hills and up for retirement (which I wish he had taken much earlier, said he didn't do keyhole surgery, but knew someone that did. I then got my referral to another OS, Mr B.

Mr B's first question, having done a quick test on leg movement and so on, was if I had had x-rays done. I hadn't, so he sent me off to get them done. On my return to his office he told me my diagnosis, and said that he felt the only solution, if I wanted to keep running, would be to refer me on to Mr J for a PAO. However, he wasn't sure if I would qualify because of my age, the window of opportunity being "between 20 and 45" — I had turned 46 three months earlier! (And yes, I did have the PAO with Mr J.) **Annick, 47 UK**

I actually saw five surgeons regarding my hip. The first made an initial diagnosis and referred me to two others, because he does not do arthroscopic surgery. I saw both of those surgeons plus two more. Two of the surgeons recommended PAO. One did not, but he doesn't regularly perform PAO, so that may be why he did not recommend I do that. I wanted to have confirmation of my diagnosis of a labral tear and bilateral hip dysplasia. I also wanted to be comfortable with my surgeon, and I found one I was confident in. **Alison 38 USA**

I asked for three opinions. First I saw two different chiropractors due to hip pain, (wrong idea, but I thought I would take a "natural approach" first). They both took x-rays and I even got an MRI with them, but got no results from a year of being treated and, more importantly, no diagnostics.

I then decided to see an OS, Dr R, who saw right away that I had dysplasia (from my x-ray), but wanted to check if the labrum was causing the pain. He ordered another MRI, arthrogram, and cortisone shot (all in one appointment). After his diagnosis (dysplasia + labral tear), Dr R sent me to another OS who specialized in labrum repair. Within this third group, I met the well-regarded Dr W, who will be doing arthroscopy to fix the labrum in my left hip soon; he introduced me to his colleague who specialized in PAO, Dr S.

Given the results of my third MRI (degenerating) and the cartilage being worn out to some extend, this last specialist is leaving it up to me to go through the PAO or not. I have opted for the labral repair only so far, as I feel that Dr S doesn't have sufficient experience and not enough PAO surgeries under his belt. I will review my options in the coming years, depending on the arthroscopy results. **Christelle, 38 USA**

And of course, you may *not* feel the need to get a second opinion!

I only got one opinion; I felt so confident in the very first surgeon I met, that there was no doubt about what he said; my x-rays were so bad that it was quite obvious even to a layperson. And I was right — he was the best possible surgeon I could have asked for. **Kris, 47 USA**

I didn't get a second opinion - it was obvious from the x-rays and pain that I needed a replacement of some kind, it was just a question of which prosthesis. The surgeon was saying all the right things and was approachable, open, with a great reputation, so I didn't feel the need for a second opinion.

Freja, 45 UK

Changing doctors

What do you do when your doctor or surgeon really doesn't have a clue? Some hippies are even told they just have to "Learn to live with the pain", by doctors and surgeons who don't know what is wrong, dismiss their patient's concerns, and aren't interested in finding out more.

Even worse are the doctors who belittle you, make you feel stupid for asking questions, are certain they're always right ("It's just a trapped nerve"), and make you doubt your own sanity (Is the pain really that bad? Maybe I'm imagining it all?). You're not imagining it. You don't have to live with this pain and disability. And you don't have to put up with being treated so badly.

I've mailed my arthrograms and x-rays to five doctors, nationwide. I've seen four doctors locally at major university-related hospitals. The only doctor to whom I've gone back is one who seems truly vested in helping me get out of pain. He's ironically the one who stands nothing to gain immediately, because I'm not yet a candidate for total hip replacement, which is all he does, surgically-speaking. He's cautioned me strongly to make surgery the last resort. He's also provided me with treatment (steroid injections and injections of Euflexxa®), and my pain has been reduced.

By reducing my pain, that doctor has bought me enough time to keep working to figure out what to have done surgically and who I should get to do it. I'm currently waiting to see specialists in New York and Boston. That doctor and I have found that Zipsor greatly reduced my pain, but my insurance wouldn't cover it, but he's gone to great lengths to get samples for me.

183

Maybe some surgeons aren't the most obviously caring people in the world. I am generalizing here, but I am not going to let a person do such huge surgeries on my hips until I see that he cares about me. I've had my films seen by about nine physicians now. I'm waiting to see two more, and then I'll decide whose opinion is right. I've even discussed with my endocrinologist what the surgeons are telling me, because I know he genuinely cares about me, so I trust how he responds when I tell him what the surgeons say. **Frances, 38 USA**

While you *do* need to respect and listen carefully what a surgeon tells you, if you really feel that you're not being heard, aren't getting satisfactory answers to your questions, and you have serious concerns about the state of your hips and mobility, then seek other opinions. True, you may *not* have hip dysplasia, you need to keep that possibility in mind — but you do need to be confident in whatever the final diagnosis is and what treatment options are proposed.

I was strung along by a doctor for over a year, who insisted that my torn labrums should not be causing me any pain and that all I had was tendonitis. At the time I was hoping he was right — that some physical therapy, rest, and stretching would take care of it and surgery would not be necessary. After I returned to him multiple times over 15 months, he should have referred me to a specialist who could have better diagnosed me.

In retrospect, I feel very angry that his incompetence led to an extra year of pain and no answers — and further deterioration of my cartilage. Now I'm on the right track, seeking out the best treatment option for me, but I could have been doing all of this research/consulting last year, before my pain became so debilitating that I have become dependent on the goodwill, time, and compassion of other people in my life to drive me to doctor appointments, fly with me to specialists in other states, pick up groceries for me, and so on. **Ivy, 43 USA**

Your original doctor doesn't need to know that you're looking further afield (although it is polite to let them know you're seeking a second opinion). It is common for patients to seek second and even third opinions, especially when dealing with uncommon conditions.

If you feel very upset with your current doctor, and that they are really not open to discussion, you can just cancel your appointments, and don't make any new ones. That's all it takes!

I asked my GP, who I had been seeing since I was about 15, if I might have something similar to my mum (who had a hip replaced at 49 due to avascular necrosis), and she totally dismissed the idea, without even getting x-rays. She is the one who sent me off to physiotherapy, which did not work, and then to a back specialist, which also did not work. It took me nearly 10 years to get a correct diagnosis. I was and am very disappointed that she didn't take me more seriously. I stopped seeing her. **Lea-Anne, 40 Australia**

As mentioned above, it can be very helpful asking for current recommendations on online groups and researching online. There is a wealth of experience out there, and groups such as Hip Chicks and HipWomen have the latest information on which surgeons are available, and what they're like.

You are not crazy. Nobody should have to live in this much pain. Do not give up. Travel to other states. Learn absolutely everything you can about the anatomy of the hip. Use the internet and find your own surgeon. Look at who does the research. Some doctors don't do as much research, but love their patients and are gifted at what they do. Sure, it's a pain, but with the internet, we have no excuses for not knowing all about our bodies, especially the parts that hurt! If someone isn't listening, trust your gut that he/she doesn't care, and keep searching for the person who does care. You have to be your own advocate. **Frances, 38 USA**

The first surgeon I saw made me feel quite stupid, and that I was totally over-exaggerating my condition and symptoms. This surgeon thought I just had a muscle tear based on a physical examination, even though a radiologist had confirmed from an x-ray that I had dysplasia! The surgeon had not even looked at my x-rays, or my notes before our appointment. Also, during the consultation the surgeon took several telephone calls. I was very upset after this first consultation, especially as I had paid for it (the NHS wait was four months, and paying was a wait of two weeks). I was very confused at this stage, because I had only just found out about dysplasia. My experiences since then, thankfully, have all been much better.

Luckily the first surgeon referred me to another surgeon, but not for the dysplasia, for a muscle tear (which I didn't have). The next surgeon I saw was fantastic, explained everything to me, and referred me to a surgeon in London for a PAO. **Olivia, 25 UK**

When you find a great new doctor or surgeon, and you've had at least one appointment with them so you're sure, you can apply to have your records transferred to the new doctor. This generally just involves sending in a letter or making a phone call to the receptionist. Some practices charge a small fee for photocopying your records for the doctor you're moving to, and there may be a form to fill in.

You may want to respectfully speak with your first doctor about it, but you honestly don't have to, if the thought fills you with dread. Cut yourself some slack, there's no need to add to your stress levels by having a heart-to-heart with them, you've got enough on your plate as it is! It's quite common for patients to change doctors, so it shouldn't be a problem.

We hope this chapter has given you some ideas on how you can be proactive on your hip journey, and how to get better medical support. In the next chapter we get down to the nitty gritty with preparing for (gulp) major hip surgery.

Chapter 11
Preparing for Surgery

Having surgery for hip dysplasia — whether a PAO, resurfacing, hip replacement, or other procedure — is a big deal. While we can't hold your hand through all the preparations (as much as we'd love to!), we hope that this chapter helps with the sometimes anxious months leading up to your operation. We cover what might happen with your surgeon and hospital, as well as what you can do to prepare before you head in for surgery.

Information sessions

In some hospital systems, you may be required to attend an information session at the hospital about your surgery. This isn't DDH-specific, so most patients will be older than you and having hip replacement for more "standard" reasons like age-related osteoarthritis. This session may include talks by a nurse from the joint replacement ward, a physiotherapist, occupational therapist, and other specialists.

They will explain the general procedures carried out at that particular hospital, so you know what to expect. They may well cover everything from the day of admission and what happens during and after surgery, to their policy on visitors, and helping you be ready to cope at home after discharge. It's a good time to ask any questions you have about admission and your stay in hospital. Make a list of questions in your medical notebook (see Chapter 10). Either take notes yourself, or bring a friend or family member who can take notes for you.

For my first THR, I volunteered for, and attended an educational class taught by the nurse, which was very informative. **Cathy, 39 USA**

It was compulsory to attend the joint replacement information session at my hospital. If I hadn't, I would have been removed from the surgery waiting list! It was useful, and gave me a chance to ask questions. I wasn't the youngest (as I'd been expecting) — a teenage girl with rheumatoid arthritis and a young man with a degenerative joint disease beat me there. **Denise, 47 Australia**

I had a pre-surgery seminar for my PAO. This was in Guys Hospital, London. A physio spoke at the seminar. I think the head nurse for orthopaedics was meant to do it, but she had been held up at a meeting. The information session wasn't compulsory, but they highly recommended it.

I did find it very useful, in terms of knowing what to expect during the stay in hospital (my experience was horrendous, however). They went over the procedure again, which was good for some people, but I knew what to expect there. It was very informative in terms of physio info and what exercises to do, and they gave out an information booklet. The physio talked about medication, what aids you might need at home, physiotherapy after surgery when you're home, and what clothes to wear.

It gave me an opportunity to ask questions before the day. The surgeon had also written out a sheet stating day by day what to expect in hospital, how you will expect to feel, what medication you'll be on, drips, catheter, and who to expect to visit you — nurse, pain team, doctors and so on. **Sian, 29 UK**

Many hospitals, however, don't offer such sessions:

There was no pre-op info session at my hospital— a serious shortcoming on their part. Two years after my resurfacing surgery, as a result of much pressure, they began a pre-op education session (non-compulsory, but strongly recommended). It was okay, but much of the info was based on what *they* wanted *you* to do (bring in all your meds, be on time etc.), and not much that was good for reassuring the nervous patient. In any case, the hospital closed it down after six months, because it was too staff-hungry for them. Instead, they now have a DVD that is given to everybody. It is not perfect, but it is fairly good, and better than nothing. **Freja, 45 UK**

I certainly didn't have any information sessions, or even receive a booklet about my PAO operation, or what to expect. I'd found out what I needed to know from my OS or from the internet. They didn't even have a PAO booklet, so they gave me a THR information booklet instead, which they admitted was ridiculous. **Sandra, 39 Australia**

There are endless differences between countries and hospital systems. Every hospital, it seems, has a different approach to this. Sometimes you only get detailed information when you see your surgeon.

My surgeon gave me some information on the PAO procedure itself before I decided to have a PAO, and in my opinion it was not written for patients, so I researched on the internet, and that's how I found out about the HipWomen group. **Lea-Anne, 40 Australia**

I attended the previous visits to the doctor and he explained to me all the procedures that I was going to undergo, but it was not a scheduled formal information session at the hospital. I just got information as it came from the doctor. **Alvaro, 38 Spain**

Preparing physically

There are quite a few things you can do in the months leading up to your surgery to be in the best shape possible.

EXERCISE

Ask your physiotherapist if they have any exercises they recommend for strengthening before surgery. Learning strengthening exercises before surgery can make them easier to do post-op, but we appreciate

> HIP TIP: Don't forget to look after your poor old "good" leg, it's taking a heavy burden as your "bad" leg deteriorates. It's common to develop temporary problems with your good leg, such as knee pain, as you limp and favour your bad side.

prior to your surgery you may be in a lot of pain, and have limited range of motion, so this may be too hard for you. Improving your upper body strength helps when you're using crutches or a wheelchair. **Isometric** exercises can help to conserve the strength of the muscles around your hip and in your legs. Exercises that improve your "core stability" muscles are also good, and can help make rehabilitation after your surgery easier.

Any exercise that improves your cardiovascular fitness will help in coping with surgery and anaesthesia. Swimming, using a stationary bike, and seated exercise programs are good non-impact aerobic exercises you can try. Using crutches is also a great workout!

If you're over your most comfortable weight, losing even a little weight will also help matters, as it reduces the stress on your joints. Try cutting back on some of the sugars and fats you normally eat, eating some more vegetables and wholegrains, cutting back on "sometimes" foods like chips, lollies, alcohol and soft drinks, and doing more exercise. We discuss this in more detail in Chapter 5.

SMOKING

If you smoke, it's very important that you do your utmost to stop at least six weeks before your surgery. Smoking significantly raises the risk of complications following elective orthopaedic surgery, specifically cardiopulmonary problems, wound complications, and the risk of requiring further surgery[66, 67]. To give you an idea of how important it is to stop smoking, here are some figures for wound complications (including infection):

In a study of almost 900 patients undergoing elective hip or knee surgery the incidence of wound complications in smokers was 23%, compared to 8% in non-smokers. These results were statistically significant[65, 67]. There is evidence that smoking cessation schemes prior to surgery are effective in helping people to abstain from smoking before surgery, and longer term[66, 68]. Our advice is, if you are a smoker, and you know you are having surgery, then get help in stopping. It is so important for your health (and that's not looking at any of the longer term effects of smoking!).

BLOOD DONATION

Blood donation practices vary from country to country. Check what the practices are where you live. The main risks of receiving blood from an unrelated donor are receiving the wrong blood type, transfusion reactions, or infection. If you live in the USA or Australia, discuss with your surgeon whether to donate your own blood (***autologous donation***) before surgery; this is not practiced in the UK.

If you can't bear the thought of donating blood, there are other safe options available. Some surgeons use a "cell saver" — this is a special machine that is used during surgery, and recycles any blood you lose, filters it, and then immediately returns it to you. This method means that you are only receiving your own blood, and the risks associated with having blood from an unrelated donor do not exist. It is also usually acceptable for Jehovah's Witnesses. Unrelated blood donations in the UK are subject to extensive donor checks and post-donation testing, which ensures that the risks of infection transmission are extremely low (1 in 30 million for hepatitis C, and 1 in 5 million for HIV)[69]. You should discuss the chances of needing a blood transfusion with your surgeon.

Following my first PAO I just had blood through the cell saver, however following my second PAO, I also required three units of blood, which I received from an unrelated donor through the UK's National Blood Service. It made such a difference to how I felt. I am extremely grateful to all those who donate their blood selflessly to help people like me. **Sophie, 29 UK**

GENERAL HEALTH

It's vital that any infections or other medical problems, like gum disease or bladder problems, are cleared up before you go in for surgery. If you have *any* bacterial infection, even something as minor as a urinary tract infection, your surgery could be postponed. The risk of bacteria getting into your hip is just too great. See your GP if you need antibiotics, or aren't sure about whether something is a problem or not. At this stage it's better to be sure!

I developed an infected ingrown toenail a week or so before my hip replacement — the first time I'd had one in years, what are the chances?! I saw my GP immediately, and was put on antibiotics. That fixed the problem, so my operation went ahead as planned. **Denise, 47 Australia**

IMMUNISATIONS

Hospitals are a great place to pick up infections of all sorts — all those sick people in one place! It's important to have your vaccinations up to date, a good month or two before your surgery, to give your body time to develop immunity. Immunisation is a proven, safe, and effective method of protection from viruses and other infectious agents. If nothing else, ensure that you have the latest flu vaccine.

Supplements

These supplements have been proven to be helpful in some cases:

Iron can be helpful, especially if you are prone to being anaemic or have heavy periods, or are donating your own blood before surgery. Note, however, that studies show that iron supplements are *not* helpful in improving recovery post-operatively[70].

Vitamin D and Calcium, when taken together, have been proven to significantly reduce fracture risk , but only if you have osteoporosis[71].

If you live in a country where exposure to sunlight is not frequent or very seasonal, then it is a significant possibility you might be vitamin D deficient and therefore benefit from this supplement. It is important to have this verified through a blood test, though.

Other Vitamins and Supplements: A supplement or vitamin is absorbed by the body only if the body needs it. If you have a low level of iron or vitamin C, for example, then taking a supplement will probably benefit you and bring these levels back to normal. However, once your levels are normal, the body will not absorb any extra and just excrete any extra that is being taken in, and not needed. As a rule, if you are healthy and eat a well-balanced diet, you do not require any supplements. If you are concerned about anything specific discuss it with your family doctor.

SUPPLEMENTS TO AVOID BEFORE SURGERY

It's easy to think that herbal supplements and vitamins are natural, and therefore safe and gentle. Unfortunately, this isn't the case. Many have active ingredients that can cause unexpected and serious problems when you have surgery.

Recent studies have reported more problems with haemorrhaging or blood clots during or after surgery, cardiovascular problems, endocrine effects, and other problems in patients who were taking herbal supplements at the time of their surgery [72, 73].

Make sure you tell your anaesthetist and surgeon *everything* that you take in detail, and be prepared to stop taking them for at least two weeks before your surgery. Only restart them after discussion with your doctor.

This is not a minor issue — your health is at stake, and it's vital that your anaesthetist knows as much as possible, so they can keep you safe. The last thing you want is something unexpected happening during surgery, because of a supplement the anaesthetist didn't know about.

The main supplements to avoid before surgery are:

◊ **Fish oils** — bleeding problems; stop at least a week before surgery.

◊ **Garlic** (capsules) — hypotension (low blood pressure), bleeding problems.

◊ **Ginkgo biloba** — bleeding problems.

◊ **Ginseng** — bleeding problems, hypoglycemia (low blood sugar).

Your medical history

Gather your medical information before your pre-admission clinic:

◊ Other doctors you currently see and their contact details.

◊ A list of medications you take regularly and their doses, including anti-depressants, contraceptives, and pain killers.

◊ A full list of any vitamins and supplements you take, even things like garlic capsules. Some of these can interfere with anaesthetics or affect blood clotting; it's important to let your anaesthetist know.

193

◊ Any other conditions you have, such as diabetes, heart problems, asthma, or hepatitis.

◊ Any dietary restrictions.

◊ Any allergies or bad reactions you've had to medications or anaesthetics in the past.

◊ Your health insurance details if you're privately insured, including any pre-authorisation codes you have been given. Make sure your insurer is aware of your upcoming procedure.

> HIP TIP: Make notes of your questions as they occur to you in a notebook in the weeks leading up to your pre-admission clinic, so you can remember them on the day. For example: "Do I need to shave my pubic area before surgery?"

Pre–admission clinic

This is generally held a few weeks or days before your surgery. During this appointment you may see some or all of: your surgeon or one of his residents (junior surgeons), your anaesthetist, a senior nurse from the joint replacement or orthopaedic ward, a physiotherapist, and an occupational therapist. You will probably have blood tests, a urine test, your height and weight measured, and your medical history reviewed. You may have an **electrocardiogram** (ECG or EKG) done, which is a tracing of your heart's activity. You might have a hip x-ray, if you haven't had any recent ones done.

Some hospitals give you antiseptic soap or wipes to use on your hip in the few days leading up to your surgery, but many don't do this — don't worry if you're not given any. Sterility is top priority in any operating theatre, and each hospital has its own standard procedures. You may be measured for DVT stockings (those lovely tight stockings that guard against blood clots).

The antiseptic was in a sponge and it was like Betadine®. I had to wash my whole body with it the night before and in the morning before surgery. I was told it was to keep me as clean as possible but they would also go over it in the theatre just before surgery. **Lea-Anne, 40 Australia**

I had to wash with a special cleanser the night before and the morning of my hip replacement. It was called Hibiclens®. The instructions given me by my surgeon's office were to wash my entire body with the Hibiclens® the night before, and then to wash my entire op side the morning of surgery. I was told it was needed to ensure that the area was as bacteria-free as possible prior to my entering the hospital. **Kris, 47 USA**

I used the antibacterial wash before my operations. It was a runny clear liquid that came in a plastic disposable tube, the type you snap the end off to open. The morning of my operation I had to wash my entire body with it. I was even provided with an instruction sheet (complete with lovely diagrams), showing which areas to pay special attention to!   And I had to make sure not to use any creams, deodorants, or skin products afterwards. It didn't smell too bad and was simple to use, although I was paranoid of contaminating myself and watched what I touched afterwards! **Sandra, 39 Australia**

I wasn't told to use any special soap, but I used Hibiclens® anyway, just because I read some women were told to prior to their PAOs. I just lathered it on from waist to mid-thigh. It stinks, but nothing horrible. **Lara, 40 USA**

The pre-admission clicnic is the time to ask more questions, and mention any concerns you have. If you have PTSD or any other serious anxieties or other conditions, now is the time to discuss them, and make sure they are written down on your records.

Stopping regular medications

Surgeons generally require that you stop some medications a few weeks before surgery. Anti-inflammatories are commonly stopped, including those found in topical gels or creams, as well as aspirin; this is because they can increase the risk of bleeding.

If you are taking warfarin, this has to be stopped prior to surgery and changed to a blood thinning injection. You will be given specific instructions with regards to what to do at your pre-admission appointment.

Some drugs are okay to continue right up to surgery; each surgeon has their own protocols on what they want you to stop taking, so discuss it with them.

CONTRACEPTION BEFORE SURGERY

Discuss this with your partner and your surgeon, as protocols vary between hospitals and specialists. The important factor to weigh up here is the risk of getting pregnant prior to surgery, and the risks anaesthetics and surgery have on a fetus, compared with the risks of using contraception.

Barrier methods such as condoms, cervical caps, and diaphragms are fine, and convey no risks to surgery. They also protect from sexually transmitted diseases.

The combined oral contraceptive pill (containing oestrogen and progesterone) does carry an increased risk of developing blood clots with major surgery. It should be stopped at least four weeks prior to elective orthopaedic surgery, and alternative methods used. Other types of contraception that just contain progesterone (progesterone-only pill, injections, or implants), and intrauterine devices are okay to use in the run up to surgery[74].

Mind you, with your increasing hip pain and disability, sex may be the last thing on the agenda, so abstinence may be all you need!

PAIN CONTROL BEFORE SURGERY

Stopping your favourite pain killers can be a right royal pain, literally! If nothing else, it will remind you of why you need the surgery in the first place. Here are some ideas for getting through this (hopefully brief) time:

◊ Ice packs — use them with a wet towel between your skin and the ice, and limit them to no longer than 20 minutes at a time, three times a day at most.

◊ Heat packs — rice or wheat packs are easily heated up in a microwave oven. Be careful not to burn yourself!

◊ Use paracetemol-based pain medications like acetaminophen (Tylenol) and paracetamol (Panadol), but not aspirin. Codeine and opiod-based medications are also usually fine, but beware of side effects like constipation.

◊ Get more rest. Stay off your feet.

◊ Avoid physically demanding activities like lots of walking, driving, or using stairs.

◊ Use raised seats to make it easier to stand up. Avoid really low sofas, chairs, beanbags, stools, and car seats.

◊ Use crutches around home, and when out.

◊ Use a wheelchair in public areas, if they're available.

◊ You might need some time off work or studies just before your surgery, if the pain is too crippling.

Disability aids

You can do quite a lot of things to get ready for your upcoming surgery. Many of these are also useful for coping with your hips in the months or even years before surgery. If you're in Australia, you may like to visit one of the Independent Living Centres, where occupational therapists can show you a vast range of disability aids, and you can try them out to find which one suits you best. In other countries, you should be able to get advice from the occupational therapists at your hospital.

I spent quite a bit of time getting the house set up: moving furniture, rugs (so the walker wouldn't get stuck). I made my room a sanctuary for rest. I know most people won't do that, but for me it really helped. Candles, low lighting, pillows, pillows, pillows, bed tray, vintage cloth napkins, vintage china, lush soft blankets — I think you get the idea. Bring a little Versailles to the bedroom! **Ina, 41 USA**

CRUTCHES

While hospitals may lend you crutches when you're discharged, it is good to have your own set. Crutches are often useful in the months before surgery, especially if you've got arthritis and are having a THR. There can be times in the months after surgery when you'll occasionally need them again — if you have a big day out when you are on your feet a lot, for example.

There are several types of crutches. A common type are the *underarm or axillary crutches* (Figure 11.1). They aren't the most comfortable of crutches. It's easy to get chafing under the arms. They can be cumbersome to use, and fall over easily.

If you're using these underarm type of crutches, you might like to securely tie a long shoelace or cord to each crutch, and drape the lace around the back of your neck and in front of your shoulders. This way you can let go of a crutch, and it won't fall over, but hang off the string.

Elbow or forearm crutches are another option, and are better for frequent use (Figure 11.2). They are also known Lofstrands, Canadian crutches, or Walk Easy™ crutches. They have a cuff which clips loosely around your forearm, and moulded hand-rests.

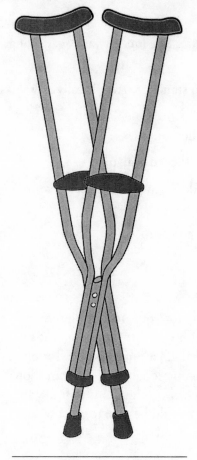

Figure 11.1: Underarm crutches.

Elbow crutches are very popular amongst the HipWomen. They're easier on your arms, there's no chafing, and they allow you to use your hands when standing still, as they clip onto your forearms and don't fall over so much. You can even turn them upside down, when you're sitting down, and use them to scrape things on the floor closer to you. You can use a single crutch, too, once you're more steady on your feet.

A word of caution about these crutches, though, from Margaret:

I've often wondered how I'd cope if I fell using elbow crutches and now I know. They are really risky to fall with as you can't release your arms from them quickly. I'm normally so careful, and must have let my guard down for a second when I didn't notice the slight step going into a shop.

The result for me was that I landed on my right side and the right handle bashed into my hip, giving me a huge bruise and lump, but worse was the cuff part and my rib cage collided, causing quite a lot of damage, including a broken rib. **Margaret, 60 UK**

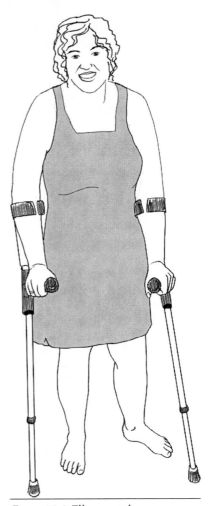

Millennial™ *crutches* (USA) are similar to elbow crutches, but are "sprung", with springs in the crutches, down near the ends. They have ergonomic handgrips, and fold up for storage. They are very popular amongst hippies.

Crutch Tips and Tricks

◊ It's a really good idea to learn to use crutches *before* your surgery, even if you don't need them yet. It's so much easier to learn when you know how your legs move, rather than after, when movement can be a bit haphazard.

Figure 11.2 Elbow crutches

◊ Using crutches is great exercise!

◊ If you live in an area with snow and ice, and your surgery is around winter, you can buy special "ice-gripper" tips for your walking stick and crutches that are slip-resistant. Look for "ice-grip tip" and "snow safety" crutches, walking sticks and accessories.

◊ Cycling or weight-lifting gloves with good palm padding can help to make using crutches a bit less painful, if your hands are suffering from the pressure.

◊ Some businesses make painted and padded crutches and walking sticks, hand-carved sticks, and other colourful accessories like crutch bags and covers. LemonAid Crutches™, CrutchBuddies™, StyleStick®, Verko, and SwitchSticks are just a few examples.

◊ Go to Chapter 19 for some crutch and walking stick crafts!

Using Crutches and Walking Sticks

The general guidelines for using crutches and walking sticks are:

◊ The hand-rests should be at wrist height when you're standing up.

◊ Your elbow should be a bit bent when holding the walking stick or crutch.

◊ Use a walking stick on your "good" side. If your left hip is the painful one, hold the walking stick in your right hand.

◊ The walking stick and your injured leg should swing forward, and strike the ground, at the same time.

◊ When stepping down a step with crutches, lead with your "bad" leg (*Down to Hell*).

◊ When stepping up a step with crutches, lead with your "good" leg (*Up to Heaven*).

Your physiotherapist or occupational therapist will also show you how to use your walking aids correctly.

WHEELCHAIRS AND WALKERS

You may find it useful to rent or buy a wheelchair or walker for a little while, if your hips deteriorate a lot. Using wheelchairs provided by shopping malls and other public venues can help a lot too.

I decided to buy a wheelchair as I knew that I would be having two PAOs over a period of about a year, and the cost of renting one for this time was more than buying my own. I intend to sell it on after I no longer need it. **Sophie, 29 UK**

I would highly recommend renting a wheelchair and using it when going out during the non-weight bearing weeks. I rented a wheelchair for a month from a local medical supply place, and I'm so happy I did. I kept it in the trunk of my car, so that I wouldn't be temped to use it in the house and become dependent on it. Of course, you need a friend to get the wheelchair out of the trunk on shopping trips. **Melanie, 29 USA**

Walkers are a type of mobility assistance device that has a walking frame with wheels and a seat, and they generally come with a basket. They are very useful when out shopping. It's almost impossible to go shopping alone on crutches, as you can't carry anything in your hands. A walker is a great option if you need the support of crutches, but don't have anyone who can be another set of hands for you.

WALKING STICK

A walking stick is a less restrictive option than crutches, and an easy prop to take with you when out and about. Generally you graduate from crutches to a walking stick, some weeks after surgery. See the previous page for tips and tricks. Foldable walking sticks are available, so you can pack it away if you don't need it.

Walking sticks have another use (as do crutches) — they are a *very* useful visual cue. Using a stick can mean that people won't hassle you when you use a disabled parking spot, jostle you at the shops when you're walking slowly, or make snide comments when you're doing slow careful exercises in the pool. At school, work, and home, people will forget about your hip. Using a stick — even at times when you could possibly manage without it — really makes a point.

CUSHIONS

You'll need these if you're having a hip replacement, as you will need to sit high, with an open angle between your body and leg (over 90°) for some weeks after surgery. You can buy wedge-shaped cushions, and solid foam blocks to raise the level of the seats around your house. Some people like to rent or buy a special chair with a high seat.

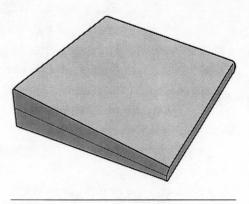

Figure 11.3: A foam wedge cushion.

LONG–HANDLED TOOLS

There are a host of long-handled tools out there, but the main ones to get are a grabber or two (pincers on a handle), a long-handled shoe horn, and a long-handled brush or sponge for the shower. Many chemists stock these sorts of supplies, and they aren't too expensive.

If you can find a long-handled razor, or razor holder, it's a great help if you like to shave your legs, and don't want to have to rely on your mum or partner to shave them for you!

If you're a keen gardener, you may like to look at some long-handled garden tools.

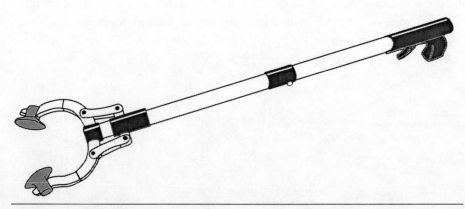

Figure 11.4: A long-handled 'grabber' tool.

'OVER–BED' TABLES

Over-bed tables make life so much easier if you're spending a lot of time in bed or sitting in chairs. Some retailers like Ikea have inexpensive ones — often sold as 'laptop computer tables'. You may also be able to rent or borrow an over-bed table from your hospital or local disability supplies shop.

BATHROOM NECESSITIES

You will probably need a raised shower chair, especially if you're having your hip replaced. It's likely that you'll be able to borrow or rent this from your hospital.

Soap-on-a-rope, or liquid soap in a pump pack, will help avoid soapy mishaps, as you won't be able to bend down to pick up anything you drop. Use a caddy that hangs off the shower head to hold everything you need up high. As mentioned above, a long-handled shower brush or sponge is useful too. Make sure the brush head is permanently fastened to the handle, or moulded in one piece, and not just wedged on.

A raised toilet chair is another necessity after many hip surgeries. Your hospital will probably lend this to you, or you can rent it from disability supply shops. There are several sorts, either a chair that sits over the toilet, or a clip-on raised seat that attaches directly to the toilet bowl. We find the "over the loo chair" easier to use, as the handles on each side are very handy in helping you get up again.

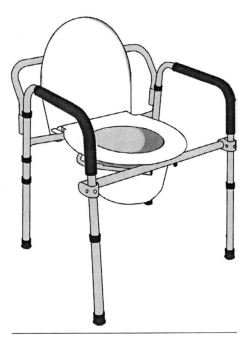

A non-slip bath mat, the rubber sort with plastic suction cups underneath, is essential, both inside the shower and on the tile floor. The last thing you want is to lose your footing and slip in the

Figure 11.5: An over-toilet chair.

bathroom. Make sure the mat is firmly attached to the floor before stepping onto it.

It's also worth asking around friends and family if anyone has these sorts of supplies — especially the shower and toilet chairs — lurking in their garages. You should only need them for three months or so.

BED LADDER

A bed ladder is a rope and dowel device that is attached to the foot of your bed, and helps you pull

Figure 11.6: Using a bed ladder.

yourself upright when in bed, by "climbing" up the rungs. It makes a massive difference in the first few weeks after surgery. They are basically a cheaper and easier to install version of an over-the-bed trapeze that is used in hospitals.

There are several models available to purchase, and we have instructions on how to make your own bed ladder in Chapter 19. Note that it only works on a bed that has very sturdy railings at the end, or legs, as the ropes have to be attached to something strong.

RAISED BED

You can purchase special sturdy lipped "cups" that sit under each bed leg. Having a higher bed is really helpful, especially after a THR, as it helps to keep that nice open angle on your hip. You can also make these changes yourself, of course, just be extra sure that the raises are very secure, and the bed can't slip off.

Raise your bed, even with bricks or wood blocks, as anything that takes the strain off your hip going from sitting/lying to upright is a boon. The plastic bag trick (sitting on a bag to swivel) can be a help getting into and up the bed. Personally I would not recommend a memory foam mattress as I found I got beached in it. Yes, it was very comfy, but when I tried to turn I was moulded into the mattress. I also couldn't get any purchase when sitting on edge of bed trying to push up from it. **Margaret, 60 UK**

SOCK GUTTERS

Sock gutters are those funny contraptions which help you to put socks on when you can't bend over. You put your sock over the end of the device, then drop it to the floor, holding onto the long straps. You then manoeuvre your foot into the opening, and by pulling on the straps, the sock is pulled up over your foot.

> **HIP TIP from a stroke-ward nurse:** To put on DVT stockings quickly and easily, roll the top half of the stocking down over the lower half. Put your foot into the foot part, and then simply pull up the top half over your calf. It really is that easy! Don't bunch it up like you would do with a sock or a stocking.

It's a good idea to try these before you buy, as it's almost impossible to tell just by looking at them which design you'll like best. Australia has Independent Living Centres, where an occupational therapist can help you try a variety of aids, such as this.

Some sock gutters are of flexible plastic, and others are frames covered with terry cloth (see Figure 11.7). We like the latter best of all, as they grip the sock really well.

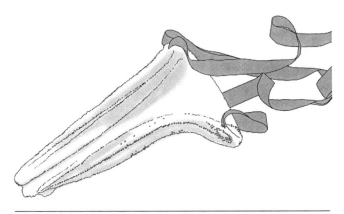

Figure 11.7: A Soxon Sock Aid, covered with terry cloth.

A variety of sock gutter is designed for putting on DVT stockings, those fashionable tight compression stockings that you're likely to be wearing after surgery. If you only need to wear them briefly in hospital, don't bother about getting an applicator.

However, if you get a blood clot after surgery, you're likely to need to wear DVT stockings for months, and these devices can be hugely helpful, as the stockings are difficult to put on otherwise. The Australian Ezy-As™ Compression Stocking/Garment Applicator is a very good design.

SLIP–ON SHOES

You're likely to have trouble putting on shoes in the weeks and months after hip surgery, so make sure you've got something that's secure on your foot, and relatively easy to slip into. A non-slip sole is vitally important, too! Your long-handled shoe horn will help you to put them on.

CLOTHES

Make sure you have some loose fitting clothes to wear at home when you get back from hospital. You won't want anything tight against your hip, wherever the incision is. A couple of pairs of tracksuit pants or loose shorts (depending on the season), a few loose t-shirts or long-sleeved tops, and loose, comfy underpants, make life easier. You may like to buy a few outfits that are a size bigger than you normally wear. Try your local secondhand clothing shop.

> **HIP TIP:** If you like to keep your hair short, it's a good idea to have a hair cut just before you go to hospital, as it may be a few months before you can get out to the hairdressers. An alternative option is to arrange for a hairdresser to come to your house.

WALKWAYS

It's important to look through your house for possible tripping hazards — it'll be hard to lift your feet in the first few weeks when you're home from hospital! Make sure walkways are clear of loose mats, and make sure there are no cords or curled carpet edges you can trip over.

MEALS IN THE FREEZER

Cook as many meals in advance as you can — make double batches when doing dinner, and freeze the extra, or set aside some cooking sessions on a few weekends. Don't forget to label the containers with the contents, date, and brief reheating instructions if you've got non-cooking family members temporarily in charge of meals. This could be a great time to teach your family how to prepare a few simple meals, too!

Even if you only have a small freezer, it's a great feeling knowing you have three or four meals ready to go in there. Dishes such as pasta sauces (without cream), soups, curries and stews freeze well, and make for fast easy meals when you're first home from hospital. Leave any dairy products like cream, sour cream, or yoghurt out of dishes you plan to freeze. Add them after reheating the dish, just before serving.

Going solo

Home preparation is especially important if you live alone, or are a single parent. The most important thing you can do is arrange for childcare when you're in hospital (if you're a parent), and support for when you're first back at home. It's also helpful if your children are able to visit you in hospital, once you're over the worst of the ordeal.

Try your best to find someone who can either come by every day for some hours, or better still, stay with you for a week or two. You might even have someone who you're willing to stay with, instead of being at your home. This is the time to pull in all those favours! Some single parent hippies have had their mothers fly in from interstate or overseas to stay while they were in hospital, and for several month post-surgery.

If you're a single parent, try to avoid putting too much responsibility onto your child or children alone, even if they're normally very capable and helpful. Talk to the kids about what they feel comfortable doing, and think about who would be a good person to be around them for them as well as for you.

If you really don't have any close friends or family you can call on, look into local social services. Many hospitals have access to programs for isolated patients, this is generally organised through a social worker or occupational therapist. Meals on Wheels is one example of a home help program. It's best to start organising this months in advance, as these things happen slowly. As you can see from Shelley's experience that follows, it is entirely possible to manage by yourself, so if this is the situation you find yourself in, don't despair!

On a practical note, don't forget to do these before you leave for hospital:

◊ Cancel the newspaper.

◊ Cancel milk or other deliveries.

◊ Arrange for a neighbour to collect your mail for you (better than the Post Office holding mail, as you'll have trouble getting there to pick it up at first). In the USA the Postal Service can hold your mail for a set period and then deliver it to you.

◊ Pay outstanding bills.

◊ Arrange for the care of any pets.

◊ Give your houseplants a good watering.

◊ Get plenty of supplies in of non-perishable things like toilet paper, UHT milk, tinned food, cereal, pasta, rice, pet food, tea, coffee, juice, and other staples.

◊ Prepare some meals in advance, and freeze them.

I had no support at all and was on my own for 99% of my recovery. I have issues asking for help, so I told myself that I could do this on my own. I had a friend pick me up from the hospital and bring me home.

Once I was home I had to get in bed, get in the shower, get dressed, put on the stockings, and so on all by myself. I am thankful that it was that way with me, because it forced me to do things. I believe my recovery went faster than most and I was able to do things so quickly post-op because I couldn't just ask someone else to do it. So many people have a spouse or parent to help them so they wonder why at three weeks post-op they can't do things like putting on their own shoes or socks. If they were on their own, they would have to figure out a way to do it and make it work. I got very good at vacuuming by standing in one spot, and vacuuming everywhere I could reach. Then I would crutch a few steps and do that area. It really is amazing what we can do if we have no choice. **Shelley, 37 USA**

Lining up support

This can be surprisingly tricky to arrange, so it's best to start early. Everyone has busy lives, after all! Hopefully a few people in your family and circle of neighbours, friends, and / or colleagues will be happy and able to provide whatever support you need — childcare, housework, cooking and cleaning, or driving you to appointments.

Sometimes your family and friends won't really "get it" about your hip problems. Occasionally, people can be quite unsympathetic and uninterested in what you're going through, let alone the need for some practical support. We discuss this problem in greater detail in Chapter 9.

You may find your partner thinks they can cope just fine without any outside assistance, but if you're not so sure on this count, you may need to insist on getting further help. Your recovery period isn't the time for you to be worrying about supporting *your* support team — get what *you* need in place well in advance, even if it means insisting on some things. Maybe you can point out to them that if you don't get help from *somewhere* else, they will need to do everything they normally do, *plus* everything you normally do, *plus* all the extra help you will need during recovery. Your partner does *not* need to do it all on their own!

Think as laterally as you can — the parents of your children's schoolfriends might be willing to drive your kids to or from school (or walk them to the bus-stop) if that's something you would normally do, and can't do while you're in recovery.

You might be lucky enough to have a trusted, capable friend of family member who is willing to coordinate assistance from others — or a group of "support people" who are willing to take turns checking what they can do for/with you.

209

Friends and family who live nearby might be willing to put on (and hang out) a load of washing, or bring over a casserole, do some shopping for you (while they're doing their own grocery shopping), take your dog for a walk, do some weeding, or vacuum a floor.

These are all chores that might not be a big deal for them, but will make a lot of difference for easing the burden on you and your family. People usually respond very positively when asked to do something specific, rather than a nebulous request for help.

While I was in hospital, my husband looked after the kids, along with his mother helping out while he was at work. After I got back from hospital, my mum came from interstate for three weeks to look after me and the kids so my husband could work. My husband was grateful for all the help we could get!
Lea-Anne, 40 Australia

For my first PAO, my mom stayed with us to help with the four kids. We have never been able to get along for more than a day or two, so I wasn't really surprised when we had a big fight five weeks later, and I asked her very nicely to leave. We were lucky to have friends and family help with meals and taking the kids to school for me.

My husband had no problems with my mom coming to stay with us, as it meant he could go to work. For my second PAO, he wasn't as lucky; I made him take three weeks off to help with the kids until I was back on my feet. I was *not* going to let my mom come and stay with us again. **Lara, 40 USA**

For my first PAO, my husband worked from home when I first came home from hospital, although he would be in the attic office, which was fine as far as I was concerned as it gave me independence just to mosey on down and do things that I felt capable of doing. If I was not capable, then I would call for him. I also had my 18 year old daughter around as she had just finished her first year at Uni. She had recently learned to drive so she was able to take me shopping or just get me out and about when needed. I was very pig-headed and didn't want help unless absolutely necessary.

As I had fallen on my arrival home from hospital, my husband was very concerned about me getting in and out of the shower safely, so made sure he was around when I wanted to wash. After about a week I was confident enough for him not to be there.

On my second PAO (right hip), again, my husband worked from home, but my daughter was up in London during the week at Uni so I was more reliant on my husband to take me places. This put a strain on things as I felt that I couldn't just say "I need to go and get some shopping (or whatever) — can we go shortly?" because as he was working, he needed to go when it was convenient to him (understandably). This made me feel constrained in what I could do, and when I could do it.

A very good friend of mine did help out when she was able to, taking me to the supermarket and wheeling the trolley around, but again, you feel a burden, even though she certainly did not do anything to make me feel that way. I also took taxis so I didn't need to keep asking mums from school for lifts.
Annick, 47 UK

Make sure you have plenty of help to rotate the burden. It is stressful to have only one person taking care of you in addition to all their other duties, work, caring for the animals, the house... Don't be afraid to ask for help prior to the surgery, maybe getting a schedule of people who can assist with your care ahead of time. **Serena, 39 USA**

Very important — don't be proud. Ask for, and accept, help when you need it.
Margaret, 60 UK

I got a lot of moral support, but only from family and very close friends, because some other friends couldn't understand the magnitude of the surgery, so I decided to keep them away until I was able to see them again. I know this sound crazy but sometimes it is better to stay away from ignorant people who will ask stupid questions.

I was able to fly my mother over from El Salvador, so she was the one that stayed with my all of my recovery time. My wife had to work and go to school at the same time, so she only saw me at night and on the weekends.
Juan, 34 USA

One thing we did that I would highly recommend is the day before we were to leave for the hospital (interstate), we threw a Send Off Party for Bri (15). She invited whoever she wanted, and my husband and I invited friends who had seen us through this.

I made sure everything we needed for the trip was all done the day before the party, and then on that day, all we did was have some fun with friends. It was open house style. We had so many well wishers — it made us all feel a little less alone. We had a lot of fun and it kept our minds off of what was ahead of us. We looked at pictures from that party the whole time we were away. And because we had done it right before we left, Bri was on everyone's minds and they all knew exactly when the surgery was. Everyone checked in frequently.
Cynthia, 47 USA

I was very lucky as my boyfriend was unemployed at the time, so was able to help me at home. In fact, he made the supports to raise the height of our bed and sofa as Occupational Health didn't have anything for modern furniture!

My boyfriend visited me every day in hospital, from 11am to 8pm. I was spoilt, I know, but I really did appreciate it. Friends and family also came to visit which was nice for both me and my boyfriend. My boyfriend is a very sociable person, and it was good for him to have someone else to talk to.

I think my boyfriend was in a bit of shock for the first couple of days I was home. He seemed to have no idea of the level of help I needed. I found this odd, as he had visited me every day in hospital! I don't supposed it helped that I was emotional, for no reason other than being relieved to be at home after being in hospital for 13 days. I phoned my best friend, who called my boyfriend, and the "problem" was easily resolved. After a few days we got into a routine which seemed to work.

Ten days after I came out of hospital, my mum fell seriously ill and was admitted to hospital. Suddenly I could no longer be the patient, Mum was. This was a difficult time for both me and my boyfriend, he was supposed to be looking for a job, but struggled to spend much time doing this, as I wanted to go up the hospital every other day. Six weeks after I had left hospital, my mum died.

In the lead up to Christmas, my best friend took me out each week to do a little shopping, have lunch, go to the cinema, and so on. This was wonderful for both me and my boyfriend. My boyfriend was able to concentrate on looking for a job, which he had struggled to do over the previous few months. **Dani, 42, UK**

I had lots of support from family and friends. My mom stayed with me for three weeks, and then my sister stayed with me for three weeks. I live on the 4th floor, and couldn't carry anything up or down the stairs while I was on crutches. When you're the strong one in the family and you have surgery, many people have a hard time. It doesn't always bring out the best in people. My girlfriend at the time fought with my mom constantly, because they both wanted to be the primary caretaker. It was frustrating for me because I didn't like needing the help, and didn't like people fighting. **Meghan, 28 USA**

I had my husband, one friend, and family come to see me; each person took turns staying at night in the hospital with me. Some friends provided meals and some support after my first, second, and third surgeries when I returned home. My extended family — aunts, cousins, parents — took me to their home to care for me since I required so much care (I was in a body brace due to the muscle reattachment) and my husband worked full time.

I found that as time went on and my condition did not improve, some friends seemed to just forget, and did not call or visit. I am going on for two years on crutches now and three hip surgeries. **Serena, 39 USA**

My mom came to live with us for several months. My man was working out of town, so it worked out for us. It was still hard living with my mother for both of us. For my next surgeries, I will figure something else out. Plus, at the time we were living in the mountains, and not close to friends or family. **Ina, 41 USA**

I had wonderful support from my parents and in-laws for all of my operations. And with three young children I needed it! The children needed to be cared for full time for the various weeks I was in hospital (two PAOs and one THR over the years). To make things easier, I arranged for my surgeries to be performed during school holidays. Then, when I was released from hospital, I stayed with my mother so she could look after the children *and* me! She had her hands full with the four of us, but it meant I could be with my children, and be cared for by my mother as well. Having our own business meant my husband wasn't able to take time off work to help. After staying at Mum's for a week, the children and I moved back home, and from then on my parents and in-laws came over every other day to help out around the house. Being on crutches for many weeks meant I needed help with housework, grocery shopping, driving to follow up appointments and physio, and so on.

I am really lucky to have family who live locally, and were willing and able to help. My husband certainly didn't have any problems at all with these arrangements — it meant all he had to do was visit the children and me, and fend for himself at home. He most likely enjoyed the time to himself!
Sandra, 39 Australia

I am very lucky that I had great support from my partner and my mum. I booked them an apartment to share close to the hospital. Luckily they were able to stay there the whole time I was in hospital. I really needed their help and support too — emotional support, bringing me things, food, sorting out the TV to watch, and helping me get up, wash, and use the commode. I went back to my parents' place after the hospital stay, as my partner's house was being refurbished. My parents are retired so it was easier to be with them and I had a room close to the toilet so it wasn't far to walk. It was very weird for me having my mum and my partner helping me to use the toilet! **Sian, 29 UK**

Arranging time off

You need to sort out your workload and leave from work or studies well in advance. It's important to give yourself at least a few days off before surgery, and several weeks and often months afterwards. We discuss the typical recovery times for each of the major hip surgeries in Chapter 4. We also discuss workplace negotiations in more detail in Chapter 9. If you're a student, you may need to defer your studies for a term, or schedule your surgery for the very start of the summer holidays. Avoid adding to the stress of your recovery by adding in assignments and exams, if at all possible.

If you are self-employed, make sure your regular clients or customers know that you won't be available for at least a month. You may be fine to get back to work earlier, but it's always best to give yourself the benefit of the doubt.

Finish outstanding projects, fill orders, get your billing up to date, pay any invoices. I you have ongoing commitments, do as much work in advance as possible. Don't forget to change your business phone message to say the business is closed until whenever. You may be able to set a 'out-of-office' message on your email as well.

Out and about after a THR

There are two major difficulties when you venture out from home after hip replacement surgery, in particular, as you're not allowed to sit on anything low, below your knee height, for several months. You'll suddenly discover just how *low* most seating is, especially in shopping centres. And that toilets are very low, too. In many places, the disabled toilets are no better, even lower than regular toilets. The next few items can help with these problems.

INFLATABLE PORTABLE PILLOW

These are available from camping and luggage shops, look for something that is easy to deflate and inflate. You can pop it in your bag or backpack, and whip it out if the need arises. Don't end up waiting, propped up on crutches, doped on pain meds, and unable to sit down for ages while your better half does the grocery shopping. Ask us how we know …

FEMALE URINATION DEVICES

Guys definitely have the upper hand here! There are a variety of ingenious devices that allow women to pee standing up. They are designed to be portable, and are either disposable or easily washed. Make sure you practice with them at home (in the shower!) well in advance of your surgery, as it takes a while to get the knack without 'leaks'.

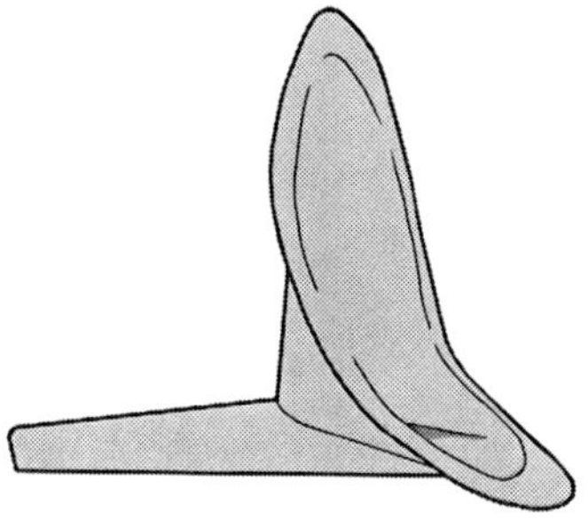

Figure 11.8: A female urination device, the Whiz Freedom®

Some brands to look for online are:

My SweetPee™, Whizzy™, P-Mate™, pStyle, SheWee®, Whiz Freedom®, Freshette, Uri-Femme, and GoGirl™

INCONTINENCE PADS

While it can be a bit embarrassing to have to buy them, incontinence pads can be a great help if you're going to be out and about, and aren't sure if the toilet seats are going to be too low (especially a problem at shopping malls, restaurants, and many other public places). One pad can hold several cups of urine, typically, so using these can extend the time you can manage being away from your home toilet.

We hope that this chapter has helped to give you some practical ideas of what you can do well in advance of your DDH surgery, to be well prepared at home. In Chapter 12, we tackle what to pack for your hospital stay — and what to leave behind!

Chapter 12
Packing for Hospital

In this chapter we reveal the most and *least* essential things to pack in your hospital bag! The combined wisdom of dozens of patients is collected here for you.

The most important things to bring

A SMALL NOTEBOOK AND PEN

This is one of the most useful and important things to bring along. You can use it to :

◊ Write down what doctors tell you on their rounds (trust us, it's hard to remember when you're on a lot of pain medications and recovering from the anaesthetics).

◊ Note down what procedures are done to you, details of any complications or medical advice, and what medication you're on.

◊ Make diary entries for each day. Track pain levels.

◊ Vent, rant, and complain in private — whether it's a terse nurse or an impossible room-mate, you can get it out of your system in your notebook.

I took a small notebook with me, and tried to write in it once an hour, or whenever I remembered. It gave me a focus, especially when I was bored, and is very interesting to look back on. I wrote thoughts, feelings, impressions, what I saw and heard. I even tried to draw a bit of what I saw, but the drawings are terrible (I blame the drugs)! **Freja, 45 UK**

EAR PLUGS AND A SLEEPING MASK

Hospitals are noisy places, that run 24 hours a day. They are not restful. These small items can make all the difference in helping you get some peace and quiet.

The most essential item that I had to bring was my eye mask. The hospital supplied pretty much everything else. The eye mask helped me block out the light and get rest any time I needed it. **Melanie, 29 USA**

MP3 PLAYER / RADIO AND HEADPHONES

Bring along your MP3 player loaded with your favourite music, podcasts, and audio books — or a small portable radio, if you prefer. This is soothing when you're feeling anxious or alone, and entertaining if you don't feel up to reading. Music can also help drown out background noise if ear plugs alone aren't enough.

Don't bring your most expensive top of the range headphones, just some comfortable cheap ones will do — it'll be less of a problem if they get broken or lost.

SOFT TOY

Yes, yes, we're all adults here, but you'd be surprised at how comforting it can be at 3 am to cuddle a teddy bear or squishy kitty, when you're in pain and alone in a dark hospital room.

I really wanted a soft toy when I got out of surgery, and didn't have one! **Lea-Anne, 40 Australia**

INSTANT HAND SANITISER GEL

Most hospitals nowadays provide hand sanitiser, but visitors don't always use it. Bring along your own small bottle, to keep by your bed. Use it as often as possible — after going to the bathroom, after eating, after seeing friends, after using crutches … you get the idea. Encourage visitors to use it as well. Just being in hospital is your biggest infection risk, so this is worth taking seriously.

MOISTURISER AND LIP BALM

Air conditioning in hospital can take its toll … these will help. If for any reason you're on oxygen, the air flow can dry out your lips and skin, and you will be very glad of some lip balm.

CLOTHES

You'll probably be in a hospital gown for the first day or so, but getting into your own clothes does feel great when you're able to! It's generally a good idea to wait until your catheter is removed before worrying about this.

Men's nightshirts, short nighties and big t-shirts work well. Remember that the nurses will need to get to your hip regularly to check and change dressings, and tend to your catheter.

Draw-string waists are important on any shorts or long pants. Baggy shorts and track suit pants work well, for when you're up and about on crutches.

Some people like to have satin pyjama bottoms to make it easier to turn sideways on the bed, but they aren't essential; many people don't like the way the fabric feels.

It won't be comfortable having anything sitting across your hips. Many of hippies "go commando" (no underwear) post-op, even for several weeks, depending on the positions of our incisions. There's no need to be embarrassed, just remember that the nurses have seen it all before! You can also bring some underwear that's a few sizes bigger than your regular size. Boxer shorts can also be more comfortable than bikini style underwear.

It's nice to be as independent as possible when you're in hospital; having to rely on a nurse to pull your underpants up for you every time you go to the toilet certainly doesn't help matters! Ditching your undies solves this problem. You'll be in bed, covered by a blanket most of the time anyway, and without undies on you won't get any wedgies when lying down! Definitely a bonus!

I just had a nightie the whole time in hospital, PJs are too hard to put on, and too difficult as you have a catheter in. I don't think I had any underpants on for a few days though! My incision was from the front of the thigh straight up. As it's not on the bikini line, underpants didn't rub, so long as they're not tight. **Sandra, 39 Australia**

There's no need to bring extra underwear. My hip was too swollen to wear them anyway. **Melanie, 29 USA**

I bought a load of cheap, cotton pants (or knickers, as we in the UK call them), mainly as I wanted them bigger than usual, so they were easy to put on, but also so I could throw them away if I wanted. On one occasion, I got my period earlier than expected, so instead of my husband getting to wash my pants, I just threw them away— he's not squeamish, but I felt a little self-conscious, so it felt like I had a little more control by doing that. **Freja, 45 UK**

Don't bring a ton of clothes — I didn't use any of them. **Serena, 39 USA**

I brought too many different types of clothing (shorts, sweats, tanks, t-shirts, sweatshirts, rompers), that I never wore. They took up a lot of space. I didn't care what I was wearing as long as I was comfortable. Stick to the basics, boxers or sweats (one size larger), and t-shirts. **Jill, 20 USA**

I did bring clothes and didn't wear them — changing was not easy to do after my PAO, so I just decided to stick with the hospital gowns. **Arpine, 28 USA**

If your hospital offers hydrotherapy during your stay, then you'll need to bring swimmers as well. Bikini style with ties on the side can be more comfortable than a one-piece suit.

SENSIBLE SOFT SHOES

You won't need shoes a lot, but they're useful for walking outside of the ward, a hobble in the hospital gardens, and for coming home. Make sure any shoes you bring are skid-proof, with a good sole.

Slippers that are not skid-proof and that don't completely encase your foot like a shoe are *not* recommended, as they can make you slip over and fall.

LONG–HANDLED SHOE HORN

This simple tool is extremely useful, not just for putting on your shoes, but to push things, pull things (if it has a bent-over end), and generally extend your reach from bed.

GRABBER

This helps no end, especially when you're mostly in bed and can't bend over. Write your name on it! See Chapter 11, and Figure 11.4.

TOILETRIES

You only really need the bare essentials — shampoo, toothbrush, toothpaste, and a brush or comb, and razor for men. If you can find it, dry shampoo is very handy in the first few days post-op — or make your own! There's a recipe in Chapter 19.

Choose shampoo and toothpaste containers with flip lids, that can easily be opened and shut with one hand, and whose lids won't fall onto the floor. Screw lids can be difficult ot manage when you're on crutches and have only got one hand free.

You can bring some basic cosmetics to help yourself feel more presentable after the first few days, but keep it simple. Some lip gloss, or an eyeliner, or whatever your must-have item is.

It is possible that the surgery will affect your hormonal cycle in hospital, so you may have an unexpected "out of cycle" period. Bring a few menstrual pads or tampons with you. The hospital can also provide these, if you get caught out.

A LITTLE MONEY

For that happy day when you finally manage to limp to the hospital café on crutches, or feel well enough to read the newspaper!

EDIBLE TREATMENTS

Throat lozenges or mints can help with your post-operative sore throat Dried apricots and prunes are good laxatives.

NECK BAG

A soft cloth bag with long handles that you can drape over your nexck is very useful — you can put things in it while you hobble about on your crutches (which won't leave you any free hands). There are instructions in Chapter 19 on how to knit your own neck bag.

The most essential thing for me was a soft, cheap reusable cotton shopping bag, with long handles enough to fit over my head and hang round my neck. Invaluable for carrying your washbag to the bathroom when on crutches!
Freja, 45 UK

MOBILE PHONE, DVD PLAYER, LAPTOP COMPUTER

You will generally be told by the hospital to leave valuables at home, because of the risk of them being stolen. However, most hippies we know have taken their mobile phones, and small laptop computers or DVD players in to hospital without any problems. You won't be roaming far from your bed for most of your stay, so you can keep an eye on your valuables. If you bring a wireless internet dongle you can even have internet access from your bed.

If you balk at taking in expensive gear, consider getting, borrowing or renting a small solid state device like an Eee PC or a portable DVD player, which will allow you to watch DVDs in bed. Just be aware that you are responsible for the safety of your gear. Don't forget to bring a few of your favourite DVDs, chargers, and headphones too!

For my teenage daughter, a phone with texting ability was essential. She never lost contact with her friends and kept everyone up to date at all times. An iPad would have been great, too. With the new version, you can use Skype.
Cynthia, 47 USA

I didn't really use my iPod. I watched films on a DVD player when I was feeling better, but music, no. Songs were just either too upbeat, or too depressing! It was hard to get into reading anything — I was so emotional and wired it was difficult to find something to hook my interest. I did however have great fun watching a box set of a surreal hospital comedy (*Green Wing*) on my DVD player.
Freja, 45 UK

I took all the usual things into hospital with me, e.g. books, music, DVDs, but found that I simply couldn't concentrate on anything. I think this was partly due to the fact that I wasn't used to having so many people around me. So in the end the only thing I did was watch TV. But at least I discovered *NCIS*!
Tina G, 41 UK

My own pillow and mobile phone were essential! Also a portable DVD player, and DVDs if you are in hospital for longer than expected.
Lea-Anne, 40 Australia

During my hospital stay I updated my blog, posted on Hip Chicks and went on Facebook. I also watched hours and hours of *NCIS* and Disney movies.
Jill, 20 USA

COMPUTER GAMES / HANDHELD CONSOLES

These can be a great way to distract yourself and while away the hours. The same proviso goes for these as for mobile phones and computers — their security is up to you. Don't forget their chargers, and earphones so the sound effects don't bother other patients in your ward.

Extras to consider

FAVOURITE FOODS

Hospital food varies, but in general, it's — well — *hospital food*. You can bring in your own treats — soft drinks/squash, special tea bags, instant coffee, Vegemite, Marmite, Nutella, chips, lollies, chocolate — whatever will help get you through the bad times! Mints, crystallised ginger, and peppermint tea can be good if you're suffering with nausea.

ICE PACK

Some hospitals may struggle to get you an ice pack quickly. Ice packs can help enormously with things like nerve pain and post-operative pain. Having your own pack (the gel type), which a nurse can pop into the freezer for you, can make life a bit easier. Bring a cloth cover for it, or a face washer (washcloth) to wrap around it. Remember that it's best if the cloth is wet. Don't leave the pack on your skin for longer than 20 minutes at a time, no more than three times a day.

MOIST DISPOSABLE WIPES

Useful for staying clean, freshening up, cleaning sticky fingers, wiping noses, and cleaning up spills.

AIR FRESHENER

To have a break from that hospital smell. This is also highly recommended if you are in a ward with other patients, especially if any of them are having digestive upsets! Be sure to check with a nurse first whether any of the other patients in your ward have asthma or chemical sensitivities first, though.

SIMPLE SOOTHING AND MINDLESS CRAFT PROJECT

Later on during your hospital stay you may be up to something easy. This is the time to work on a garter stitch scarf, for example, but not a Fair Isle beanie.

Bri (15) just used her phone during her hospital stay, to text people. Everything else we took to hospital was used by me to help me pass the time sitting with her. She slept a lot. I spent a lot of time decorating her crutches. It helped to pass the time and I felt like I was being useful to her, when there really wasn't anything I could do for her. **Cynthia, 47 USA**

When I was in hospital for my THR, all I could manage craftwise was knitting a baby beanie, just mindless plain knitting round, and round, and round. **Denise, 47 Australia**

IPAD OR SIMILAR DEVICE

Truly wonderful devices, but you will be responsible for their safety in the hospital. Make sure you bring the charger and headphones as well!

My iPad was the best thing to bring. I listened to music, watched a movie or two, and had the internet to communicate with friends via email. That and my Blackberry were the best things because I felt in touch with the outside world, and my friends and family could reach out to me. I was alone at times between visits from hubby and family, so being in touch with others was helpful to my spirits. **Alison, 38 USA**

BOOKS, MAGAZINES, AND PUZZLE BOOKS

Stick to simple stuff. You may find that you're too exhausted or out of it to feel up to reading, but you never know. Just bring in one or two books or magazines; you can always ask family or friends to bring in more reading material later if you need it. Many patients find they are simply too tired to want to read, and don't have the mental concentration to manage it. Some people, however, do enjoy the hospital time to catch up on their reading!

I was finally able to read a lot — probably more during the two weeks post-surgery than the whole pre-surgery year. **Arpine, 28 USA**

I brought several books and some puzzles with me, but didn't look at a single one of them during my whole nine day stay. I didn't even listen to audio books or watch much TV either — I didn't seem to have the ability to follow anything. Mostly all I did was sleep, talk with visitors, and listen to music. **Denise, 47 Australia**

I couldn't read, too much pain/itching/nausea from meds. I brought books, but didn't open a single one. **Alison, 38 USA**

I brought a few books and never touched them; the meds they gave me made my vision blurry for a long time. I did … nothing at all. Really didn't have the energy or desire. **Kris, 47 USA**

I could hardly read a page; so better leave books at home, because you can´t concentrate on them. I spoke a lot with my wife, and relatives came to see me. **Alvaro, 38 Spain**

I was too weak after my first PAO to do anything at all. I was able to do *light* reading, like magazines, but not books, after my second PAO. What little energy I had was spent on visitors, and talking on the phone. **Lara, 40 USA**

What to leave behind

COMPLICATED ACTIVITIES

Honest. You won't be able to concentrate for anything tricky. The lingering effects of the anaesthetics and the ongoing effects of the pain killers you'll be on, coupled with disrupted sleep and pain, mean your brain will be rather addled for a while. Leave the lace knitting, embroidery, computer programming, complicated novels, and cryptic crosswords for when you're back home. In fact, you can probably safely leave *all* books at home.

All I did in the hospital was sleep, attempt to eat, do physical therapy, talk with visitors, and recover. I was even too tired to listen to the audio book that I had brought with me. **Melanie, 29 USA**

SLIPPERS

Dr Sophie has seen far too may accidents on the wards cause by patients wearing slippers. If you're wearing footwear, it needs to be firmly attached to your feet, and have non-slip soles.

WORK OR SCHOOL WORK

Don't bring work of any sort to hospital. You don't need the pressure, and you won't be able to concentrate on it or do a good job of it in any case. Give yourself a break!

We travelled 1,200 miles with those darn schoolbooks. They were heavy and guilt-inducing! Bri (15) was not up to schoolwork and the books were nothing but trouble! **Cynthia, 47 USA**

JEWELLERY AND WATCH

You won't need to wear either in hospital. This doesn't apply to "immoveable" jewellery like wedding rings, and important things like medical alert bracelets or necklaces.

More suggestions and experiences

Don't take too much with you, as you won't really need it. Chances are clothes-wise you will either wear hospital gowns or nightwear (comfy and loose). Take a cardigan or dressing gown, and going home clothes. Easy to slip-on shoes, but not loose ones. I wore my trainers.

Earplugs are a must. I didn't read at all — couldn't hold any concentration. My laptop was *essential*. I just watched TV, emailed everyone on the HipWomen group, and updated my blog. I had planned to do embroidery and read a French book, but did neither — and still haven't! LOL! **Annick, 47 UK**

I didn't bring many things to the hospital, just a few lollies/sweets, and some magazines and books. I actually wasn't able to read the books, the pain medication meant I couldn't concentrate long enough to get through a paragraph. I even struggled with the magazines. Watching mindless television was about all I could muster. Of course, I had no problems with the lollies! **Sandra, 39 Australia**

Don't bother bringing magazines or reading material or dry shampoo (that can really wait until you get into a shower). I was too out of it most of the time; watching movies (in parts), and taking phone calls was about all I could manage. **Lea-Anne, 40 Australia**

I was in too much pain to barely watch TV, and then so out of it on morphine that I didn't need any entertainment. **Ina, 41 USA**

My favourite things to bring to hospital were: lightly scented body spray, baby wipes, lip balm, cell phone, and a small notebook and pen to keep track of med dosing, questions to ask, and things to remember. A small plastic tote, labeled with my name, was helpful to keep all of it on top of the bedside table within easy reach.

My sister surprised me with a beautiful photo book of my four daughters to take to the hospital with me. It was comforting and heartwarming to be able to look at my girls throughout my hospital stay. **Lara, 40 USA**

My favourite thing in hospital was a hand-knit prayer shawl from my dearest friend; the hospital was chilly and rather than have a sweatshirt or regular shawl, I brought this one. It gave me such comfort. The most essential thing I had was my iPod. **Kris, 47 USA**

I didn't really use much at the hospital except my phone so I could email people. I can see an iPad being helpful, but I didn't read, I didn't turn on the TV, I just slept. I was very drugged up while I was in the hospital, so I couldn't do much. **Meghan, 28 USA**

My personal blanket/pillow, back scratcher (epidurals and anesthesia effects can make you itchy), TV series and movies, spray bottle with water (refreshing), and my laptop/cell phone/camera. **Jill, 20 USA**

My essentials list: iPod Touch for music, games, and to listen to my audio Bible; computer and cell phone; hand cleaner wipes to clean hands after using bed pan or eating; hats for bad hair days; cough drops for sore throat and dry mouth from pain medicine. **Serena, 39 USA**

Calming Essential Oils — lavender, chamomile, clary sage blend; **Rose misting water** — I always run hot, and this helps me with stress; **Stuffed animal** — I really needed some fluffy comfort; **A neck pillow** filled with flax seed & lavender flowers —U-shaped, it provides great neck support. **Pstyle-pee funnel.** I couldn't sit for weeks to urinate. This is an amazing product. You can stand up and pee, without having to take your clothes off to squat. [See Chapter 11 for more information on these.] **Ina, 41 USA**

My thoughtful sister gave me a small, framed photo of me and my family having a nice time on holiday. It made me feel much better, plus it reminded the nurses that I was a real person, not just a procedure for them to deal with. **Freja, 45 UK**

The Bare Bones List

If you want to travel *really* light, here's our list of the bare necessities :

◊ Notebook and pen.

◊ Earplugs.

◊ Toothbrush, toothpaste, shampoo, comb; razor for men.

◊ Moisturising skin cream, lip balm.

◊ Hand sanitiser gel.

◊ One or two pyjamas (or you can wear the hospital PJs and dressing gowns the entire time if you wish).

◊ Health insurance / national medical coverage membership card.

◊ A little bit of cash for a snack or a newspaper.

Now that you're packed and ready to go, we can move on to Chapter 13 and your hospital stay!

Chapter 13
Surgery and Hospital

The operation day, day of surgery, New Hip Day, PAO-Day, the BIG Day, Oh #*%*! Day — whatever you decide to call it, this chapter helps you prepare for what might happen, and hopefully make the process a little bit less stressful.

Actual procedures vary so much between surgeons, hospitals, and countries that this can only be a rough guide. What actually happens to you during your admission and stay is likely to be slightly different, but let us say one thing: if you think you should be having something or would like something during your stay in hospital, *ask*.

Surgery day

Morning or all-day operating lists usually start around 8 am, and afternoon lists generally start around 1 pm. You can be anywhere on the surgery list, and this is not usually confirmed until the actual day. Therefore, all the patients for that list need to be there before it starts. This means that you might have a long wait on your hands on the day of surgery.

RISE AND SHINE

If you have to be at the hospital at a specific time, make sure everything is ready the day before to avoid any last minute panic and *set an alarm* (or two) that allows you plenty of time to get to the hospital. It's best if you have someone near and dear who can come in with you, so make sure they're ready on time too! If you're taking any anti-anxiety medications, such as diazepam, in the days before surgery, you will probably not be allowed to drive.

WASHING

Have a decent wash before you leave home. If you are having a PAO, THR, or other osteotomy, it may be the last shower or bath you get for a few days. Some hospitals have a policy of asking patients to use a special antiseptic antimicrobial wash or soap, such as Hibiclens®, the day before and/or the morning of admission. It may come in a prepared sponge, or as a liquid wash. This procedure seems to be more common in private hospitals, rather than public ones.

When I had my THR, I was told that I had to use Hibiscrub® for 2 days pre-surgery. This was okay, but it wasn't the most pleasant of smells and it seemed to dry out my skin. I was also quite reluctant to use it to wash my hair, and found that even weeks after my surgery I was still having to use loads more conditioner than usual. **Tina B, 36 UK**

While the nurses were prepping me for surgery, they washed my hip and upper thigh with special soap to make sure the incision site was clean. I didn't have to do this before I came in. It felt completely normal, just like they were washing me with soap and a washcloth. **Emily, 18 USA**

NIL BY MOUTH (NBM)

Nil by mouth means having absolutely *nothing* to eat or drink, not even water. Prior to surgery, you will need to be nil by mouth. There are no absolute rules with regard to how long you should not have anything to eat or drink, so follow whatever instructions you are given. In general, you will have to stop eating anywhere from six to 12 hours prior to the procedure.

NBM is to ensure there is nothing in your stomach prior to a general anaesthetic or sedation. It reduces the risk of any stomach contents over-spilling into your lungs, which can cause pneumonia and breathing problems. Many places will ask you to be NBM for all food and drink from 6 hours prior to the start of the operating list. So, unless you are first on the list, you are likely to hungry and thirsty for a significant period of time!

If you have regular medication you're still on before surgery, you should be allowed to take this with a tiny sip of water. Check with your doctor.

DON'T BE LATE

Do not be late, it may mean you miss your slot on the operating list. If you think there might be problems with traffic or public transport, or you have to travel at peak hour, anticipate them and plan your journey. It's better to be early!

Exactly what happens when you arrive at hospital varies depending on the hospital. You may be asked to wait in a clinic before your operation, or you may be allocated to a bed. If you are allocated to a bed or room, it is easier to arrange your belongings and get set up. However, if you are asked to wait in a clinic setting, your bed is often not allocated to you until after your operation. Make sure you have packed what you think you might need immediately after your operation (not a lot!), and put it near the top of your bag, or give it to a relative or friend to look after.

If you have another serious medical condition, or live a long distance from the hospital, you may be admitted to the hospital (or occasionally a hotel) the night before. Some hospitals admit you the day before if your operation is happening early the next day. It really depends on each hospital, how many beds they have free, and their standard policies.

VALUABLES

Try not to take any valuables in to hospital on your surgery day. Most hospitals will get you to sign a disclaimer for all your belongings. If you've decided to bring things like a laptop or portable DVD player (see Chapter 12), ask your family or friends bring these in to you after a day or two. You really won't be up to using them until then.

Some hospitals will look after your bags and walking stick/crutches, adding identification labels and locking them up while you're in theatre, and delivering them to your room afterwards. If this service isn't offered, entrust your bags and walking aids to a friend or relative to look after. Do not leave them in your room (because it won't be secure), or in bags that will be left unattended.

WHEN YOU ARRIVE

You will be seen by a nurse or health care assistant, who will check all your personal details, put a wrist band (listing your name, birth date, and any allergies) on you, and check all your observations (temperature, heart rate, blood pressure, and oxygen saturation).

If you are having trouble with fear or PTSD, make sure that any plans established during your pre-admission clinic are set in motion — you may be allowed to have a mild sedative tablet when you arrive, and be able to wait in a room separate from the public waiting room if you're upset. Remember to ask, they'll be only too happy to help you out. If you know you'll be a mess, it helps to have an advocate there with you — your partner or a close friend — who can arrange things for you.

Fear

If you are not even the tiniest bit nervous before an operation, you're either not being honest, or there is something wrong! Due to a multitude of reasons, some people deal with their nerves or fear better than others. It is a good idea to plan for dealing with fear.

There are quite a few good distraction techniques, some of which we cover in Chapter 7. It is very helpful having a loved one with you while you are waiting to go to theatre. They can help to distract you (a game of cards, anyone?), be your advocate if you're freaking out, or give you a shoulder massage. Do whatever works to keep your mind off things, whether it's reading, listening to music, solving sudoku, talking to your buddy, knitting, audio books, or playing games on your mobile phone.

If you know that this period is going to be particularly difficult for you — again, plan for it. Hopefully you've had some sort of pre-admission appointment. Make sure you mention your fears then. There are several possibilities for medication (e.g. diazepam) that can be given to you to calm you down. Remember to ask for them, or get your family doctor or GP to prescribe them for you so you have them available on the day.

I knew I was going to be terrified on the morning of my surgery (I have PTSD around surgery). I coped fairly well until I had to sign the admission papers, but then I couldn't stop weeping in abject fear. I was allowed to wait in a private room with my husband, instead of sitting in the public waiting room with everyone staring at me. After I'd changed into the hospital garb I was given a temazepam sedative tablet — something I'd arranged for during my pre-admission clinic. Once I'd calmed down, I was allowed to lie on a trolley bed, to the side of the main waiting area, and listen to music on my iPod, with my husband by my side. That really helped a lot. **Denise, 47 Australia**

Getting Changed

There are two main possibilities here, depending on the hospital policy. If you are already allocated a room or bed, then you will likely change into a gown, some paper underwear and slippers (stylish, hey!) and be taken to the operating theatre from there, either walking, in a wheelchair, or on a hospital bed. If you are waiting in a clinic setting, you will be taken to a holding room near the theatres, and get changed into the same attire there. Your clothes will be stored and returned to you after the surgery.

Pre-operative procedures

SEEING THE SURGICAL TEAM

Someone from the team (usually this is the surgeon, but can sometimes be one of the other doctors who works for them) will come and see you. If you haven't signed a consent form prior to this, they will go through one with you. Even if you have signed one at a prior consultation, it is a good idea to ask them to go through it again, especially the risks and complications so that you have everything clear in your mind.

You will then be "marked". This means that the side you are being operated on is marked in indelible marker. This is reassuring for you and is standard policy in every hospital — you will not be allowed near the operating theatres without that mark on you. You don't want the wrong hip being operated on, after all!

QUESTIONS

This is your last opportunity to ask the surgeon any last minute questions. It's best to write this list before the day of the surgery, as we can guarantee you will forget things in all the stress and anxiety that leads up to the procedure. Please remember — there is no such thing as a stupid question — if you want to know the answer, then it is worth asking!

Anaesthetics

SEEING THE ANAESTHETIST

The anaesthetic doctor is the one who will be looking after your breathing, vital signs, pain relief, and anaesthesia, while the surgeon does the operation. They will discuss the different types and combinations of anaesthetic that are used and what the best option is for you. Sometimes this discussion is held during your pre-admission clinic.

Anaesthesia can be broadly categorised into 3 main sections:

GENERAL ANAESTHETICS

This is the type of anaesthesia where you are completely put to sleep, usually through an intravenous (IV) line in the back of your hand or arm. All you will know is one moment you are in the anaesthetic room breathing some oxygen through a mask, and the next, you are in recovery and it is all over. Most hip surgery is done under **general anaesthesia**.

REGIONAL ANAESTHETICS

Regional anaesthesia numbs a "region" of your body, so you cannot feel anything in that area. For the purposes of hip surgery, this is done via either an epidural or a spinal anaesthetic. This involves having a small needle or cannula put into your back. Then medication is given through it to block all the signals that the nerves receive from your lower body (including pain). It is the same option that is used for either pain relief or a Caesarean during child birth. This is sometimes done in addition to a general anaesthetic. If the spinal anaesthetic is used without a general, then you are also given sedation to relax you during the procedure.

My surgeon typically uses a spinal anaesthesia for PAOs (rather than a general), which was my first choice. I chose this option because it would be faster coming out of sedation, less nausea afterwards, and I felt, overall, less risky.

I'm glad I had a spinal. I remember absolutely nothing after they wiped my back with a cold swab in the operating room right before surgery, until "waking up" in the recovery room afterwards. The effects of the spinal lasted until the next morning, when I began feeling mild pain only. That was well controlled with Oxycontin twice a day, and Dilaudid when needed.

I think the only con was that my blood pressure was very low coming out of surgery, and I think the spinal may have contributed to that. **Lara, 40 USA**

While I was not awake for my hip replacement, I was supposed to be. Up until the morning of my surgery, the plan was for an epidural with twilight sedation. When I met with the anesthesiologist in the pre-op area, he and his assistant strongly suggested general anesthesia instead — he said that he felt it was the better option for the surgery. It really wasn't what I needed to hear 30 minutes prior to my surgery (!) but I went with his suggestion. I am glad I changed my mind about the anesthesia, though, even if it did come at the 11th hour. **Kris, 47 USA**

LOCAL ANAESTHESIA

This sort of anaesthesia is where a much smaller area is made numb; it is never used in isolation in hip surgery. Local anaesthetics are used prior to an epidural insertion and near the end of an operation to numb the incision sites, so you are in less pain when you wake up. A femoral nerve block is a type of local anaesthesia which affects the femoral nerve, which makes the thigh numb.

Your anaesthetist is also involved in pain relief post-operatively. If you have an epidural or femoral nerve block, it can be continued to keep you "numb" after surgery for the first 24–48 hours. An epidural can offer complete pain relief, but prevents you from moving easily, which is why it is only used for a short period, to get you through the worst.

If you've had a lot of childhood hip surgeries, it's possible that you may not be a candidate for a femoral nerve block, because of scar tissue and/or deformed anatomy in that region.

Surgery

Surgey is covered in detail in Chapter 4, but on the day — what will you be aware of?

After getting changed, you will be taken into the anaesthetic room, which is a small room next to the operating room. Here you will be attached to monitoring equipment, and an IV may be put in here. If you need to go to the toilet, don't be ashamed to ask, the nurse will happily get you a bed pan. When they are ready for you, the type of anaesthetic you're having will be administered. You may be in this room for as long as an hour.

If you are having a general anaesthetic, you will have no memory of the operating room itself. If you are having a regional anaesthesia with sedation, then you may have some hazy memories, however the drugs they give you as a sedative usually affect your memory, so you might not remember anything. If you are not asleep, you might be allowed to listen to some music. It's worth asking if you can take an MP3 player and small headphones in with you if you're going to be conscious.

You can discuss your options for anaesthetics with the anaesthetist, if you think you would prefer a certain method just say so, they should take this into account.

RECOVERY ROOM

After the anaesthetic room, this is the next place you are likely to have any working memory of, and you will have probably been there at least two to three times longer than you think.

You will have one-on-one care here; the nurses will look after you until you're awake, comfortable, and stable enough to go to the ward. You'll have at least one, sometimes two, IV lines into your hands/arms; these shouldn't hurt. They're there so pain relief and fluids can be given to you quickly and easily. Your blood pressure will be checked regularly, and you'll have a little machine clipped (painlessly) on your finger or ear lobe that records the amount of oxygen in your blood and your heart rate. You will also be wearing an oxygen mask — which you may well try to pull off! It's not uncommon to vomit during recovery, as a side effect from the anaesthetics. You may or may not remember this.

Other things you may have will depend on the type of surgery and can include:

◊ **A drain** around the incision site (you shouldn't be able to feel this).

◊ **Epidural lines**.

◊ **A urinary catheter**. This keeps your bladder empty and means you do not have to go to the bathroom (although you may not like the sound of this, if you have had a PAO or THR you will really appreciate it. It's put in while you are asleep in theatre).

◊ **A foam wedge** (*abduction pillow*) between your legs to stop them from accidentally crossing.

Depending on your surgeon's preferences, you will either have some form of dressing over the incision site or a type of skin glue.

If you are thirsty, say so and ask for a drink, you should be allowed to have some water.

If you are in pain, say so, they will be able to give you something — ***do not*** suffer without saying anything.

If you cannot feel your legs, don't panic! The most likely scenario is you're still being affected by regional anaesthetic; if you tell the nurse looking after you, they can reassure you.

INTENSIVE CARE

This section within the hospital is also known as the **intensive care unit** (ICU), intensive therapy/treatment unit (ITU), or **post-anaesthetic care unit** (PACU).

The majority of patients undergoing surgery for DDH do not go to intensive care. There are a few centres in the USA that routinely send their patients there, but this is not generally the case in the UK or Australia. If you have other medical problems, you may have been told prior to the surgery that you will be going to intensive care — this is usually as a precaution. If you find yourself in intensive care unexpectedly after your surgery, do not panic; though, to be honest, it is unlikely you will panic because of all the medications you'll be on!

There will be a nurse with you all the time who can explain everything to you. Being in ITU means you will have one-on-one care (like in recovery) for longer and never left alone, which you will likely find reassuring after your surgery. To be honest, being in ITU is harder on your friends and family than it is on you, you may well have very little memory of the experience.

I have a pre-existing heart problem that meant I was due to go to ITU for 24 hours after my second PAO. Unfortunately, the surgery caused more problems than anticipated, and I was there for three days. Although I was very nervous about going to intensive care before the surgery (the fact I could be *that* sick scared me the most), I was very grateful for it. I actually have very little memory of the time I spent there. If I was told I needed to go there again I wouldn't be so worried, partly because I know I probably won't remember it! **Sophie, 29 UK**

The first night

You will be in a bed on a ward; this may be a side room, a room of your own, or in a room with other patients. It is unlikely you will get much sleep the first night for a multitude of reasons. These include:

◊ Pain.

◊ Having your blood pressure, heart rate, and oxygen levels checked every 2–4 hours (this is necessary the first night and while you are on any type of pain infusion, be it an epidural or PCA pump).

◊ Nausea.

◊ Being in a strange place, in a strange bed.

◊ Noises on the ward, monitors beeping, noise from other patients, and so on.

If you are on a PCA pump (**patient-controlled analgesia**), it usually is hospital policy to keep you on oxygen as long as you are using it; this can be via a mask or two little tubes that sit at the entrance to your nose (*nasal specs*). Oxygen supplied this way can dry your mouth out, so make sure you can reach a drink easily and use lip balm.

Make sure you have your call button and your PCA button, if you've got one, within easy reach.

If you are in pain or feel sick, ***tell someone***, as these are both solvable problems that should be dealt with.

Try to sleep, but if that's not happening for you, just rest and try not to stress about not sleeping.

I had little pain when waking post-PAO — they kept me on the PCA pump for a couple of days. I was up the next morning for gentle physiotherapy, learning how to walk on the crutches, and how to get in and out of bed. I was at London Clinic, and so started hydrotherapy the following afternoon, once the catheter was taken out. **Annick, 47 UK**

Hospital routine

We just have to mention one important thing here: despite being in bed or a chair most of the time and moving very little, being in hospital is not restful, not even close.

The lights on the ward generally go on between 6 and 7 am, and often in this period all the blood tests for the day are done so that the results are ready for the doctors during the day. What a way to be woken up! The nurses will do drug rounds (when all the prescribed drugs for patients on the ward are given out) at set times during the day, the first in the morning usually occurs between 7 and 8 am, and the last one in the evening is around 10 pm. The lights will usually be turned off after this.

DOCTORS' ROUNDS

These usually occur in the mornings, but can happen at any point during the day. You should expect to be seen by a doctor once a day, however this may not always be your surgeon, but one of his team. Remember to take notes of what your doctor says in your notebook (see Chapter 12).

NURSING

It's important to be realistic about what to expect in hospital. Hospital isn't a hotel (although, sadly, you're likely to see other patients who have this attitude). Your nurses have been well trained in how to care for you, but they are only human, and they have to manage a great many patients.

It's wonderful if you're in a specialist joint replacement ward with nurses especially trained in caring for joint replacement patients. However, you may be in a general orthopaedic or surgical ward, with nurses who don't know much about PAOs, for example. If you know more about hip dysplasia than your nurse, don't be judgemental — you have a rare condition, after all. As long as you're being cared for competently and kindly, it doesn't matter whether the nursing staff know a lot about DDH.

It is reasonable to expect to always be treated with respect and kindness, like a person, not just "The hip in Bed 3". If a nurse is truly unkind to you, by all means mention it to someone in authority. In these situations, it's really helpful to have a friend or family member acting as your advocate.

It's too much to expect you to handle conflicts when you're recovering from major surgery — ask them to carry out any negotiations for you.

That said, do keep in mind that doctors and nurses are people too, and have constant demands on their time, every single shift. There are plenty of demanding, selfish patients who make their life difficult. There are also patients who are very sick who require more input from medical staff, so if you can help redress the balance by being patient and cooperative, they will certainly appreciate it.

BED CHANGES AND WASHING

This is usually done in the mornings. If you're unable to get out of bed, nurses can change sheets from under you. This usually involves some rolling (which can be uncomfortable and painful in the first few days post-op) or getting out of bed to sit in a chair, but it is worth it to have lovely clean sheets under you.

MEALS

These occur at set times and may be quite different to when you are used to eating. It is a good idea to have some snacks of your own, in case you get hungry outside these times. You will probably be given a menu each day, so you can select what you'd like for your meals the next day. If you have special dietary needs (diabetic, vegetarian, coeliac etc) make sure you notify the hospital during your pre-admission clinic (see Chapter 11).

NIGHT TIMES

Although the lights will go out on the ward, you can have your bed light on if you cannot sleep (just keep the curtains drawn so you don't disturb other patients). Nights in hospital are noisy. If you have your own room it will be quieter, but if you're sharing a room with other patients it can take some getting used to. Ear plugs and an eye mask are invaluable in helping you to get some sleep.

QUIET TIMES

Most hospitals have a quiet time during the day where no visitors are allowed, so patients can get some rest during the day. Take advantage of this if you can, it really is peaceful during these times — though "peaceful" is a relative term in hospital! If there are no set visiting hours, you can always invent your own. Ask your advocate to tell people that you're only available during certain hours each day, to ensure you get enough rest. Visitors can be surprisingly exhausting in that first week post-op.

By the way, it's quite normal to have a dip in mood around three days after surgery, so don't be surprised if you're suddenly teary and not coping. This too will pass.

Post-op pain management

Pain medication has been discussed in detail in Chapter 3, but there are a few points about managing your pain on the wards. We won't lie to you — the first few days post-op can be very unpleasant and painful. But if you can follow our guidelines below, you will hopefully not suffer too much.

Pain is far easier to control if you stay on top of it — it is *much* harder to get good pain relief once it has got out of control. Therefore, as soon as you start to feel uncomfortable ***ask for some medication***. This is important, especially on busy wards, as it may take the nursing staff some time to get the medication to you. If you can anticipate your need it can help. Do not suffer in silence.

PATIENT–CONTROLLED ANALGESIA (PCA) AND EPIDURALS

These are both infusions, drugs delivered to you through a thin flexible intravenous tube on a special pump, and will need changing. The machines make loud beeping noises when the bag containing the medication runs out, which usually grabs someone's attention. If it looks like the machine is running low, tell a nurse as soon as you can. This means they can prepare the next infusion so that it is ready before the machine runs out, minimising the time you are without any pain relief.

A PCA pump is often used after hip surgery for a day or two. This means you have an infusion of pain relief medication, delivered through an IV, which you control. You receive a dose by pressing a button. There is a mechanism which prevents you from being able to give yourself too much, so you can never overdose.

The benefits of a PCA mean your pain relief is under *your* control, which can be good on busy wards. However, the down side is you can't press it when you are asleep, so it can be difficult to sleep for any significant period — after a few hours you might be woken up by pain. You then need to press the button again to get your pain relief under control.

Remember to discuss your options for post-op pain management with the anaesthetist. It's also important to remember that if you're in pain, you need to *ask* for pain relief.

OTHER ANALGESIA

Pain relief can also be given the traditional way at set times during the day, with extra doses that can be given when needed. These can be given orally, or with an injection, either into a muscle or a vein. The PCA or epidural will only stay up for a day or two post-op, and then you will be moved to oral pain relief — you can't really go home with lots of lines still attached to you, however well they work! IV lines also hamper your mobility, it's much harder to get back to walking when you have tubes attached to you.

It's important to take your regular pain medication even if the pain is bearable, when it is given to you. As mentioned previously, it's easier to stay on top of pain, rather than having to regain control of it once it has become unbearable.

SIDE EFFECTS

Unfortunately no drug is without these, however, whether or not you experience them, and to what degree, varies greatly from person to person. As a rule, all morphine/opiod-based drugs will cause some degree of nausea; the doctors will anticipate this and will prescribe anti-sickness medication for you as well. If you start to feel sick, the same principle applies as with pain medication; do not let it get out of control, ask for something to help with nausea early. Feeling nauseated is one of the most unpleasant feelings and is very treatable, so tell a nurse early.

245

Another extremely common side effect of strong pain killers is constipation. In fact, just expect that it's going to happen. For some patients, the pain from constipation and getting their digestive system working properly again can be worse than that from their hip surgery! However, the good news is there are things you can do so that it doesn't get out of control.

I know this is common, but post-op (PAO) I am finding the Oxycodone slow release and immediate relief tablets are making me very constipated and giving me stomach pain. I am trying to take the immediate relief ones less, but if I leave it over 10 hours between each tablets I tend to have severe gripping stomach pains and be on the loo for some time! Also very common, initially they made me very sick, but now I just feel lethargic. They also seem to affect my short term memory and my ears feel pressurised, like they are popping — or need to pop all the time, like when flying. **Sian, 29 UK**

I had Oxycodone post-op and that worked great. However at one point I was given Vicodin, due to a shortage of the Oxycodone, and had a horrific reaction to it. Uncontrollable crying, excessive sweating, and terrifying psychotic thoughts and nightmares. All from just ONE DOSE! I never took the Vicodin again. **Kris, 47 USA**

The only problem I had with taking pain medications is that they constipated me. I had a lot of problems with that in the hospital and ended up having to have three enemas. They had to switch my medication from Oxycodon (a narcotic) to Tramadol (a non-narcotic) to help with the constipation too. **Emily, 18 USA**

I definitely felt a "detox" after I stopped taking my meds. I had cold sweats and shakes for two days but then I was fine. **Alison, 38 USA**

I took Percoset right after surgery but also had to take an anti-nausea medication to tolerate it. After a week or so, I switched to Norco and did not need to take the anti-nausea medication with it. I had trouble with sleeplessness after I stopped the Norco about three weeks after surgery. It took about two or three weeks before I slept normally. **Louise, 45 USA**

Constipation Tips and Tricks

◊ Anticipate it and start doing something about it from day one post-op — do not wait. Discuss this with your medical team.

◊ Non-pharmacological methods:

- Prunes

- Rhubarb

- Dried apricots

- Liquorice

- Increase your fibre intake.

- Stay very well hydrated, drink lots of water and other drinks. Dehydration makes constipation worse.

◊ Laxatives. There are various different types, it is an idea to start with the softeners, and then if these do not work progress to the stimulant types:

- Stool softeners e.g. lactulose, which is a thick sweet syrup that you drink.

- Stool bulkers e.g. senna, a tablet or a sachet that you drink.

- Stimulants e.g. bisacodyl, this is a tablet that stimulates your bowels to move.

- Suppositories e.g. glycerol. This goes up the bottom and makes it easier to pass a stool.

- Enemas – again, these go up the bottom and are stronger than suppositories.

◊ Side effects. The laxatives that you take orally can all cause stomach cramps, the severity of which can vary from person to person (some people feel nothing).

◊ Anticipation. Once you think you *might* need to go, try to tell someone early, even if it means you are in the bathroom a while. Your mobility is hampered and you'll probably need help to reach the bathroom, so don't leave it to the last minute to ask if you can help it.

I had a truly awful time getting my intestines back to working order after my THR, despite being on all sorts of laxatives and drinking heaps of water. You know it's bad when you push the nurse call button while you're on the loo, nearly fainting from the pain! Not a way anyone wants to be seen, and by a male nurse ... I ended up having to have an enema. The whole process was very unpleasant. **Denise, 47 Australia**

Menstruation

If you have been involved in the choice of surgery date and have regular or controlled cycles, then choosing a date where you know you won't have a period during the first couple of weeks of recovery is ideal. However, many women do not have this option, so having a period while in hospital is a very real possibility. The physical stress of surgery can also muck around with your cycle, and you may find yourself having an early period.

If this does happen, please don't worry about it. The medical staff are used to dealing with this on a regular basis and, believe us, there's nothing they haven't seen before, and very often they have tricks up their sleeves that can make things easier. If you have a catheter in, pads can be used with or without underwear, depending on how comfortable you feel with wearing them (and getting them on). Hospital bathrooms are fitted with sanitary disposal units.

The important thing is to wear whatever you feel most comfortable with (pads or tampons). If you are using tampons, ensure that you're able to put them in and remove them without difficulty or too much pain.

Tell the surgeon before surgery (or a member of the team) if you are menstruating. Although you might feel embarrassed, it is important for them to know. Surgeons get concerned if they see blood in unexpected places during or after surgery!

How Embarrassing!

It can feel unbearably embarrassing when you have an x-ray or other scan, and there, for all to see, is your genital piercing, tampon, IUD, or something else "down there" which you thought was hidden! Please don't worry — the technicians and doctors are professional. They have honestly seen it all before, and will barely notice. They certainly aren't laughing at you.

One of the many problems with hip problems is, well, they're your hips, and doctors, nurses, physiotherapists and a host of other people are going to poke and prod around your hip joint, bum, and groin before, during, and after surgery. There's a distinct lack of privacy, and this can be rather galling at times.

Your medical team are truly are more concerned about the state of your joints, muscles, ligaments, and caring for your incision, than whether you've had a bikini waxing or not. If you can get into the mind set of "This is medical, and not personal", it really does help.

After being in the hospital for a week with only a small gown to cover me I got used to being exposed to my doctors, nurses, and family, and in those moments of exposure I didn't care at all. The only problem I had was during PT when my physical therapist would massage me near my groin and I was very ticklish there. **Emily, 18 USA**

Washing and dressing

WASHING

For the first few days you will probably need a significant amount of help to wash yourself. This will usually take the form of a bed sponge bath. The best thing is to let the nursing staff help you, they are experts at doing this and do their utmost to maintain your dignity.

If you are feeling embarrassed, please try not to be: firstly, the staff are extremely professional and do this regularly, and secondly, after the surgery, trust us, you really won't care. The joy of feeling clean after staying in bed for 24 hours is well worth the effort.

After a few days you will be able to have a shower. Showering in the hospital is something to really take advantage of. The bathrooms are designed for use by people after surgery. They have special chairs to sit on, lots of grab bars, call buttons if you need extra help, and are designed for wheelchairs, so you don't even have to get up to walk to the bathroom. For many people, showering in the hospital is actually easier than when they get home!

I had a male nurse who was helping wash me, and he was great! And although I was embarrassed a bit at first (he was a young guy, I am a young girl after all), I just tried not to think about it, and just be a patient. And it was great!
Arpine, 28 USA

DRESSING

This is a very personal thing. Some people prefer to remain in the hospital gowns during their stay, others prefer to bring their own nightwear, and others prefer to put on regular clothes on every day. We talk about clothes for your hospital stay in Chapter 12. Whatever you decide, there will be people to help you get dressed (and undressed). You can pick up tips from staff, and sometimes other patients, on the best way to manage things like socks and underwear. There are some dressing tips in Chapter 15 too.

Room–mates

Room-mates can make your stay in hospital more enjoyable (if enjoyable is the right word for being in hospital), or pretty miserable. Unfortunately, this is one aspect you just have to keep your fingers crossed about. If you aren't in a private room, and want some peace and quiet, pulling the curtains around your bed is a good way of sending the message that you want some privacy, and one the majority of people understand. You are likely to be a fair bit younger than most other patients in the orthopaedic ward, and practically no-one else will be in for DDH (that's how special you are!), so it can be easy to feel somewhat isolated at times.

THE GOOD

You can make friends with your room-mates; hospital can be quite lonely (despite being surrounded by people all the time), and having someone in the next bed who you can talk with makes you stay more pleasant for them and you. Being a good room-mate is just as important — as a "younger patient" you can even be helpful to the nursing staff.

The elderly woman in the bed next to mine was being prepared for her hip replacement early one morning, and she was utterly terrified, crying, and alone. The nurses didn't have time to stay with her. I was just able to get out of bed, and hobbled around on my crutches to her. I sat with her, held her hand, talked with her about my recent surgery (just two days earlier), and reassured her that everything would be fine. I lent her my iPod with some soothing classical music to listen to while she waited to be taken to theatre. It really helped her, and it made me feel good to be able to help someone. The nurses also thanked me later for helping out. **Denise, 47 Australia**

Being a Good Room-Mate Tips and Tricks

◊ The nurse call button makes a noise, so while you should definitely press it when you really need it, continuously pressing it unnecessarily can be annoying.

◊ Have all your stuff organised and close to hand, keep your area tidy.

◊ Wear headphones to listen to music or the TV.

◊ Ask your visitors to stick to visiting hours, and limit the number who are there at any one time. Your advocate can help with this if you're not up to much.

◊ Be conscious of how loud you and your visitors are, try to keep the general noise level down. Stop children from running around too much.

◊ Treat all hospital staff with courtesy, kindness, and respect, from your surgeon and nurses, to the people who deliver your meals and empty the bins.

◊ Be patient, sometimes the nurses have more urgent things to do.

◊ Talk on the phone quietly and don't receive calls after a certain time at night (that isn't too late!).

◊ If you are allowed a mobile/cell phone in hospital, keep it on silent or vibration mode all the time, so ringing and SMS tones don't bother people if you forget to turn it off at night.

THE BAD AND THE UGLY

There are lots of reasons that people fall into these categories — dementia, deafness, fear, diarrhoea, non-stop talking, visiting relatives (and whole extended clans), and just plain old difficult personalities. Some are potentially avoidable, others unfortunately not. If there is a circumstance that can be improved, either ask the patient directly (but politely) or ask the nurses, whichever you feel more comfortable doing —via the nursing staff is likely to be more successful.

Difficult Room-Mate Tips and Tricks

◊ Ear plugs to block out the noise — but do inform the nurses looking after you that you have them in!

◊ Eye mask to block the light, or to avoid being spoken to.

◊ Headphones so you can use the radio, music, or audiobooks to block out the noise, but listening to music/TV can also be relaxing and also means you are not disturbing anyone else.

◊ Air freshener. Let's hope you don't need to use it, but a life saver if you do! Check first that using it won't cause any problems for asthmatics or allergic patients (or nurses!).

◊ Keeping a blog/journal. You can complain to your heart's content there! If you are writing about your room-mates in public online, be *absolutely* sure not to give out their identities, call them "Mrs Crazy", or "Bed 2" or something like that!

◊ Ask the difficult patient's relatives to teach their family member how to use the nurses' call button — it sounds simple and obvious, but is something a surprising number of people struggle to understand.

There was a woman who was brought up from having bladder surgery, still sedated really, who came to demanding Mini Gems, Jaffa cakes, and Snickers! She also snored *so* loudly. She did ask me once if her snoring kept me awake, I said no, I never slept! She somehow seemed relieved by my response! The next day my boyfriend bought me some ear plugs.
Danielle, 42 UK

The very elderly woman with a broken collarbone in the bed opposite me was a complete nightmare. She had dementia, was incredibly rude to all the staff, complained and talked non-stop even when alone, refused food and treatment, complained when nurses were looking after other patients ("Why are you helping *her*?!" spat with contempt), and even accused the staff of molesting her. She carried on like this all day *and* all night. Even the nurses were losing their temper with her, and took away her nurse call button. The rest of us were in tears by the second night of this unrelenting bad behaviour, but the nursing staff weren't able to move her elsewhere (and yes, we did ask!).

I could only block her out by having ear plugs in and listening to loud music on headphones at the same time. I also wrote down the ridiculous things she was saying in my notebook, as something to do instead of getting upset and angry. Eventually she was put onto mild sedatives which made her miraculously pleasant and biddable. Thank goodness! **Denise, 47 Australia**

I had one terrible room-mate after my PAO. She had been in the hospital for a week after falling asleep/drunk at the wheel, and driving into a telephone pole. She broke her hip in the accident. She was constantly yelling for the nurses to help her with everything. She made the nurses come to bring her bedpans, because she refused to even try to get up and wheel to the restroom. If they were five minutes late with her pain meds, she was yelling for them and pushing the nurse call button. When they finally got there, she would yell at them for being late with the meds. She also had sleep apnea so whenever she would sleep she would stop breathing and all of her monitors would go off. That happened at least 10 times a night every night.

I dealt with it by just trying to ignore her. At night I convinced the nurse to change the settings on her oxygen supply so it wouldn't always go off all night. The funny thing is I got out of the hospital after three nights, and she was still there, refusing to leave. **Shelley, 37 USA**

Visitors

Having relatives and friends visit you in hospital is quite often the highlight of the day, and what most people look forward to. Have a close relative (be this partner/ parent/ sibling/ close friend) act as your advocate — set this up before you have your surgery, and make sure they understand what you need from them.

Having visitors, whilst being lovely, can also be *very* tiring; your descent into utter exhaustion can happen surprisingly quickly. Ask your advocate to arrange visits so that not everyone turns up on the same day, or to a certain time limit. Setting limits is important, and you're not the one to be doing it.

Having an advocate is also appreciated by the medical staff. It's time consuming and unnecessary to have the doctors explain to every single member of the family what has been happening — having one person who all your other relatives and friends go to for information is useful.

If you find yourself tired whilst you have visitors there, just explain to them that you're feeling exhausted and need to rest. You can suggest that they leave for a short period (coffee at the hospital café perhaps?), and then come back, which is a nice suggestion if they have travelled a long way to see you. This can give you time to rest.

We strongly recommend that your visitors don't bring their young children with them, if at all possible.

GIFTS

Not all hospitals/wards allow flowers or plants, so advise your visitors to avoid bringing these. Even if they are allowed, the person next to you might suffer with hayfever (or the nurses, who also don't have time to care for flowers). While flowers are lovely, save them for when you are recuperating at home.

YOUNG CHILDREN

If you want your children to visit you in hospital, you'll need to have done some preparation with them at home, and reinforce it on the days they visit. Children can find hospitals very scary, especially seeing Mummy or Daddy attached to lots of machines. They may also not understand why they cannot just jump on your lap, or why you can't just hop out of bed easily. Even teenagers can find the experience upsetting, especially if you're looking battered about and ill. Here are some quotes on how others have dealt with this:

My daughter, who was 6 at a time of my PAO, really benefitted from visiting a PT session at the hospital when they taught me to walk on crutches. It helped her understand how hard it is to go through what I was going through, and taught her a bit to be careful with me and my hips.

As far as the hospital, she was very excited to see me — she just likes doctors and all the medical stuff. So she wasn't scared or upset, but wanted to explore everything, touch my leg, lay on my bed, and was thrilled watching nurses taking blood pressure, asking me questions, and all that fun stuff that gave her new ideas on how to play doctors. **Arpine, 28 USA**

Take your children to hospital with you when you are admitted, so they see where you are going to be (not just somewhere that their imagination can make up!) and can be a bit familiar with your ward/room. Show them your wound when possible, as often as they need to look at it. It makes it real and helps them to remember to be careful of that side! Keep it real but keep it simple. **Lea-Anne, 40 Australia**

I lined up my sister and best friend plus their kids to each take two days to stay with my kids at my house while I was in the hospital. The kids had "sleepover" parties at my house, with fun activities to keep them occupied during the day. When my kids did come and visit, I had presents ready for them to distract them from how awful I must have looked. We also talked beforehand, how Mommy would have IV lines giving me water and medicines. **Lara, 40 USA**

I've had DDH surgery twice with young children. The first time was probably the hardest, as I was in for two weeks and my daughter was just 18 months old.

We made videos of me saying hello from hospital, and she enjoyed making one to send back to me. (It was very hard for me to watch it, though, as I missed her so much, but lovely at the same time!). We had a lot of help from relatives, who would bring her in and take her home after a short visit, so I could spend time with my husband.

For my second surgery, I was only in a few days and by this time I had two children. They visited for short periods, and after the initial visit, where they were not sure what to make of me, they enjoyed walking up and down the wards, and the "older" ladies (there are often lots having hip replacements) enjoyed speaking with them. My husband had to take them on lots of short walks around the hospital to the shop and garden to keep them occupied, and not too noisy on the ward. When visiting with the children, you don't get a lot of time to see your partner!

My kids wanted to sit next to me on the bed, and it was hard to explain they couldn't. I showed them my dressings and drains, which helped them to realise why it was sore. **Clare, 39 UK**

Dealing with complications

The common complications related to each individual type of surgery and how they are recognised and treated are discussed in detail in Chapter 4. However, we think it's important to mention some tips on dealing with a complication, if you are unfortunate enough to suffer with one.

There can be complications or problems that occur after surgery that no-one can anticipate, however much planning or informed consent is carried out (we both speak from experience!). All the major or common complications should have been discussed with you before your surgery.

I developed pericarditis (inflammation of the lining of the heart) after my first PAO. Neither the surgeon or my cardiologist had seen this being precipitated by surgery before — was it the surgery or just coincidence? Not something I will ever know. **Sophie, 29 UK**

In the hospital, I had to have two pints of blood transfused due to extra blood loss during the surgery. With the additional arthritis my surgeon discovered during surgery, the operation took longer sthan expected, so I had more blood loss. I was frustrated about having to use someone else's blood since I didn't have any taken prior to surgery. **Laurie, 50 USA**

A few days after surgery my oxygen levels dropped and I was having some trouble breathing. I was still pretty out of it, so wasn't aware of any of this. A slew of tests and scans later I was diagnosed with a small pulmonary embolism. I was given a higher dose of Clexane, put on warfarin (for six months), and had oxygen through a tube resting under my nose. I had to stay in hospital for a few extra days while they got my INR reading in the correct level. At least I was in hospital, and got a fast diagnosis and rapid treatment! **Denise, 47 Australia**

So, you've been told that something has gone wrong, or you've developed a complication … what next?

Well the good news is that most complications are treatable, however, there may not be a quick-fix option. For example, treatment for blood clots takes months, and nerve damage takes months, if not years, to completely heal.

Many steps are taken to reduce the incidence of complications. For example, to reduce infection, special laminar flow operating rooms are used and patients are given antibiotics during the procedure; to help avoid deep vein thrombosis, patients get to wear extremely fetching DVT compression socks and are given blood thinning injections. If, despite all this, a complication or two happens, just remember it is not your fault. In fact, with many of the complications that occur it's no-one's fault — plain old bad luck usually has a significant part to play.

However, as mentioned in Chapter 4, the surgical experience of your surgeon does play a part with relation to complications. With regard to to periacetabular osteotomies, increased surgical experience has been associated with a decrease in complications[75, 19]. The importance of choosing a surgeon who is experienced in dealing with hip dysplasia in adults cannot be emphasised enough.

We hope that this chapter has prepared you for what might happen while you are a patient in hospital, and how to make the experience as bearable as possible. The early stages of recovery and going home will be covered in the next chapter.

Chapter 14
Recovery at Home

You've made it through the surgery and your hospital stay, well done! But now the really hard work starts. We hope this chapter gives you some idea of what to expect, and how to manage with the physical and emotional rollercoaster in the first couple of weeks after major hip surgery.

Coming home from hospital is always exciting, but can be accompanied by a significant amount of anxiety and apprehension. There's a lot to think about, and take responsibility for.

Before you leave the hospital

The most important thing is to make sure you are ready to leave. By this we don't just mean that you've got all your belongings packed up (you obviously need to have this done too) but that, within yourself, you feel physically well enough to manage at home. This is may not be an option in the USA, however. It doesn't matter whether you feel ready or not. When it's time to go, you have to go, unless you have a specific medical symptom that prevents you from being discharged.

Once you are confident you feel up to leaving, and are only taking oral pain medications, make sure you've packed up the following:

◊ Toiletries from the bathroom.

◊ Chargers plugged into the wall.

◊ Clothes, shoes, and personal items.

◊ Anything you've lent to room-mates.

◊ Cards and gifts.

One of the nurses or other ward staff can usually help you pack your belongings, if a friend or relative is not around. Remember to check under the bed, in the closet, and the drawers!

DISCHARGE LETTER AND MEDICATIONS

Your discharge letter is a summary of everything that's happened to you while in hospital, including details of your surgery, which someone on your surgeon's team prepares for your GP. It is especially important, because it provides details of any new medication, and changes to your current prescriptions. It includes the results from any tests you've had in hospital, and details of any complications, if you had any. Sometimes there is information for your GP about follow up appointments that your surgeon wants you to have (physiotherapy, orthopaedic check ups, and so on).

You are generally given this letter to take to your GP in person; sometimes the hospital will post it directly to your GP. It is a good idea to see your GP soon after discharge from hospital, within a week or so, to touch base with them. You will probably need repeat prescriptions for things like antibiotics and pain relief from them soon after returning home, in any case.

It is *vital* to ensure you have enough medication to last you until you can get to your GP or family doctor for a repeat prescription. Most hospitals give you enough medications for one to two weeks. However, some hospitals just supply you with prescriptions, which you have to fill on your way home (which is another reason why it's important to have someone helping you!). Once you get home, it's important to work out how many days you have of medication, and booking an appointment with your GP early so you don't run out. If you're discharged around major public holidays (such as Easter or Christmas), make sure you take this into account.

I was given a script which the nurses had filled for me at the hospital, and this was enough for about 10 days. Then I went back to the hospital to have the staples removed, and I got another script from the surgeon which lasted about a week. Apparently they cannot do a repeat on certain pills, like tramadol and Endone. I was given 100 Panadeine Forte with three repeats and I still have some left! **Lea-Anne, 40 Australia**

No medicine was supplied by the hospital after my PAO or arthroscopies. I had to go to the pharmacy right after being discharged, which was difficult each time, when all I wanted to do was to get home and rest. **Louise, 45 USA**

I had to make sure that they called my local pharmacist to have my medication (high dose Vicodin) ready for me to pick up on the way home ... I lived an hour away and didn't want to travel more then I needed. Had I not done that, they wouldn't have had it ready for me, as it wasn't a common medication. **Ina, 41 USA**

I was discharged with two weeks' worth of medication. Unfortunately, the discharge form that went to my GP stated "No repeat prescriptions'" This meant that when my meds ran out, it took me several phone calls to get a repeat prescription. Why this happened I don't know, but it made life very stressful! **Dani, 42 UK**

PHYSIOTHERAPY AND DISABILITY AIDS

It is particularly important to have your physiotherapy/physical therapy referrals or appointments sorted out before you leave, especially if you are going to have your rehabilitation at a place other than the hospital where you had your surgery.

Make sure that all the disability aids (that you have hired or borrowed yourself, or that have been arranged by the hospital) are set up at home. Even a few days without these will make everyday life immediately post-op much harder work. If you haven't been provided with anything, it's never too late to get a few basics; see Chapter 11.

Travelling home

Where you live in relation to the hospital has significant implications for your journey home. There are a few things that are applicable, however, whatever the method.

Firstly, take some pain killers and some anti-sickness medication *before* you leave. Even if you haven't had any problems with nausea in the hospital, you may well do when travelling. It's better to anticipate this, and be prepared for it. On this note, it might be worth taking a couple of disposable sick bowls along with you.

If at all possible, we don't recommend using public transport to get home. Arrange for a family member or friend to pick you up. If you live nearby and don't have anyone to pick you up, a taxi is a worthwhile investment. After most hip surgeries, getting in and out of a car is possible —just take your time (more on this below and in Chapter 15). Bring a pillow to cushion your hip, you'll be surprised at what bits hurt when you're in a car. Going over bumps and around corners can be hard to handle in the early days.

I only had a one and a half hour car ride, but I learned after the first PAO to make sure I took my pain meds about half an hour before discharge. That means making sure your partner goes to the pharmacy to pick-up your meds an *hour* or more before you are discharged, instead of on the way home! **Lara, 40 USA**

I travelled from Scotland to London for my surgery (500 miles). I came home by train. One thing that made a difference was booking a seat in first class, as there was much more room than the normal carriages. I was also able to pre-book assistance at the train station, which meant I got a wheelchair to the train, and we got help with our bags. In my opinion it was definitely much easier to do a journey like that post-op by train, rather than by air or by car. The worst part of the journey was the five minute car ride home from the train station !
Deirdre, 36 UK

After my left PAO I went home in the car — it was about an hour's journey. We have an SUV which was quite hard to get into, as it is high up, but once in I was pretty comfortable. The most discomfort was from my wound where the seat belt went across it, so I just pulled it forward a bit so it wasn't tight on it. I also tipped the chair back slightly so I wasn't so upright as I would normally have been. Other than that, all went fine.

After my right PAO, with an undiagnosed stress fracture to my inferior pubic ramus, sitting in the car was very uncomfortable, but I got home with not too much difficulty. **Annick, 47 UK**

AIR TRAVEL

Surgeons who specialise in adult hip dysplasia are not particularly common, and you may have travelled a long way for your surgery. Consequently, flying may be the best way to get home. It's a good idea to inform the airlines that you are travelling after surgery, and what you mobility level is. Some airlines may require documentation from your doctor to say that you are safe to fly. All airports have "special assistance", where they can help you get around the airport, through security, and onto the plane. You can either request this when you book the tickets (recommended), or when you check-in. Make sure you have enough pain medication in your hand luggage. A small pillow to cushion your hip is a good idea. Remember to flex your feet during the trip to help avoid blood clots.

I traveled about 800 miles by plane from Denver to St. Louis. The surgical coordinator recommended waiting about a week after my discharge before going home to make sure there were no issues. Since I was able to stay with friends and family, there were no additional hotel costs. When I traveled home, I was on Southwest Airlines, which has open seating. I was able to pre-board and sit in the front aisle of the plane (with the operated side away from the aisle). Before going through security, a skycap at the airport arranged for a wheelchair and a pass for a friend to go to the gate with me. I had to have a pat-down, and the only thing I would have done differently was to ask for it to be done in private.

Unfortunately, my plane was late, so my friend had to leave before I boarded. That meant leaving the wheelchair at the gate, and crutching over to the bathroom with my backpack. Of course, due to security, the staff at the gate absolutely would not let me leave my bag at the gate.

Once on the plane, it all went smoothly. The airline attendant got my bag down after takeoff, and she got my crutches down later when I needed to use the bathroom. I also took a ziplock bag for ice that she also filled for me. She seemed a little put out by my requests, but she was polite -— maybe she was just tired because the plane was late. I took pain medicine on schedule, so I was as comfortable as I could be on the plane.

When I reached Denver, a skycap took me to out of the secured area to meet my brother, and he collected my bags and carried them to his car. I stayed the night at his house, and he drove me the 50 miles back home in the mountains the next morning. **Louise, 45 USA**

I flew home from Minnesota to Kentucky via commercial airline, and then had a one hour drive at the end of the flight. It was very difficult. My mom was with me — I couldn't have managed it by myself at all. We did make arrangements with the airline ahead of time, and that helped. We also had help meet us at the airport. Pain medications were essential!! We also rented our own lightweight wheelchair to take along with us. It was worth it's weight in gold! I was exhausted for a few days after. **Katie, 47 USA**

LONG CAR JOURNEYS

If you're making a very long trip home by car, consider splitting the journey into several small distances, or even over a couple of days. Remember to have a cushion or two to help get you into a comfortable position, especially when sitting for lengthy times. It's important to keep your legs moving on long journeys to avoid blood clots, so wear DVT stockings, flex your feet back and forth now and then to help avoid blood pooling in your legs, and take regular breaks when you can get out of the car and hobble around for a little bit.

It took me four hours to get home from the hospital ... I recommend driving in an SUV. Have plenty of pillows, a blanket, take pain meds to stay on top of pain, and bring an iPod for music to soothe and relax you. **Serena, 39 USA**

I had to travel from Tacoma, Washington to Beaverton, Oregon; roughly a four hour trip. I was in the back seat of our SUV and very uncomfortable. The SUV sits high, so we had trouble getting my stiff sore leg up high enough to go through the door, and then lower to go back into the foot well.

We needed to keep my right hip towards the back so that we could prop pillows without them falling to the floor, and the seats were stiff, so my mom and husband ran to a local store and bought a bunch of various sized pillows with the hopes that we could fashion a comfortable chair in the stiff SUV back seats.

We tried to prop my hip on pillows and lay the seat back, but I ended up mostly laying sideways over the seat belt through most of it. Everyone else got out at a McDonalds for a pit stop, but I was too afraid of hurting myself getting back in and out of the car, and potentially falling on the slick floor, as I didn't have a walker with me and wasn't too keen on the crutches yet. I didn't get out until we got to the house.

For the next surgery we are considering renting a van; it will be lower and have a wider door, so I will have more room to maneuver my sore hip into a comfortable position. The seats in a van can independently lay back more like a recliner, too. **Lisa, 40 USA**

I travelled eight hours by car for my PAO surgery. Pennsylvania to Boston. I really didn't experience too much discomfort. My mom and I split the trip up into two four-hour rides, and stayed with relatives. **Jill, 20 USA**

I traveled from Boston home to NY (Westchester County) — about three hours. When they wheeled me down to the car I started crying. I was scared. I went from the bubble of protection and care of the hospital to home where, although I had support of family and friends, I was fearful of something going wrong and not having my doctor near me.

The trip itself was okay. We stopped at a rest stop to go to the bathroom. It took me about half an hour to crutch from the car in a handicap spot to the bathroom, and it was very painful. I would recommend making sure you time your pain meds well so they last you through the trip, or take a top up dose mid-way so you don't get uncomfortable during the trip. I had to lean back in the seat to keep my hip at the right angle (over 90°), so a reclining seat was necessary. **Alison, 38 USA**

We travelled a long way for Bri's PAO surgery — 1,200 miles. Bri (15) was not allowed to get on a plane for five days after she was released from the hospital, so we had to go to a hotel. My advice — stay as close to the hospital as possible. We were only a block away, but it was snowing on the day of our follow up visit, and I couldn't get her up and down the hill.

She also couldn't go to PT during those days because it was very rough getting around the city in a wheelchair, and very cold weather. Make sure the hotel is handicap accessible. Most will tell you they are, but you have to ask about showers. None near our hospital had a separate shower stall, only tubs with shower heads. Bri couldn't lift her leg to get in. Dry shampoo came in handy. Also, make sure the beds are comfortable. If they are uncomfortable to you (carer), they will be worthless to the patient. We had to change rooms after the first night of no sleep.

The plane ride home for Bri was pain free. She had taken her meds and she handled it just fine. But it was grueling for me. Ideally, two adults to help would have been best. The assistance from the airline was terrible. The day we came home was really one of the worst days of my life. Again, Bri was fine — no pain. But *so* stressful for the person taking care of the patient. **Cynthia, 47 USA**

Back home

Hurrah! You're back home at last! You can sleep in your own bed, have hugs with your family, cuddles with your pets, and your favourite things to eat and drink! But these first couple of weeks immediately after surgery are still difficult. We've covered the main things that you may find challenging.

As we have mentioned in previous chapters, it's extremely unlikely that you will be able to manage at home completely independently (although it *is* possible, see Chapter 11). We won't try to cover every single aspect of recovering from major hip surgery here, as everyone's home situation is different. We have focused on some of the main aspects that we felt were important, and likely to have the most impact.

One thing to remember though — the person, or people, helping you are probably doing their best. Caring for someone is no easy task.

HIP RESTRICTIONS

If you've had a resurfacing or THR, these will be an important part of your life for a few months. It is *vital* that you adhere to them from Day One. You must keep your hip-torso angle at 90° or more, must not cross your operated leg across the midline of your body, and must not twist on your hips. To learn more about managing with hip restrictions in detail, see Chapter 15.

PAIN RELIEF

Medication
Please refer to Chapter 3 for details on specific medications and how they work.

Hopefully by the time you have got to the stage that you needed surgery for your DDH, you have a pretty good idea of which pain relief medications work well for you, and which ones don't.

One of the biggest things to be aware of with pain is that once it's out of control, it's *much* harder to get it back under control. Therefore, do your best to anticipate it, even if you feel you're managing with the pain when it is time to take another dose.

Certainly in the early post-op stages, it's better to just take the painkillers regularly, instead of waiting for the pain to get bad. If you miss doses, it will take even longer (and probably higher doses) of medication before you feel comfortable again.

The length of time you take pain killers depends on the type of surgery you have had. Pain is a very personal experience. While one person may find they can manage without pain killers early, another may require them for much longer. Some complications, such as nerve damage, can result in longer-lasting pain. Whichever situation you find yourself in, don't worry; pain is what you say it is — in other words, if you feel you're in pain, you *are*, and you need to do something about it. Don't suffer just because you feel you "should" be off drugs by a certain time. Everyone is different.

It is important to remember the non-pharmacological things that can make a difference (listed below). Despite even the best pain relief, it's possible that you will not feel comfortable for some time. All we can say is, it will get better — it just takes time.

Ice
Ice packs can help relieve swelling and soothe painful areas. Remember to keep it on for only 20 minutes, and to have the ice pack wrapped in a wet cloth so it isn't in direct contact with your skin. Chapter 3 has more information on using ice packs, and Chapter 19 has instructions on how to make an ice pack easily at home.

Heat
Heat packs can help with muscle pains or relaxing sore muscles, and are often simply nice and soothing, especially in the winter or colder months. There are instructions on how to make your own heat pack in Chapter 19.

Pillows

In bed, on sofas, on dining table chairs, pillows can make a big difference. They can be used to make a harder chair softer, allowing you to sit more comfortably to eat at a table, as sitting on hard chairs in the early recovery period is rather uncomfortable. Use them to support your operated side, or relieve the pressure on it either sitting or in bed. In bed you need to keep a pillow between your legs after THR, even when lying on your back, to make sure your operated leg doesn't move past the midline of your body. When you are allowed to lie on your side in bed, a pillow between your legs is makes it more comfortable, and is essential after hip replacement.

Recliners

Some people find that having a chair that can change position makes a big difference, and they even sleep in them in the early stages of recovery.

Before Exercise

Half an hour before your physiotherapy session, take a decent dose of pain killers, this will mean you get more out of the session. You may find the day after a physiotherapy or hydrotherapy session you're in more pain, and more exhausted, than usual. This is normal. Do not use pain medications to "push through" the pain, though, and end up doing more exercise than you can handle.

Dosset Boxes

Use a **dosset box** /pill organiser for your various medications. These are boxes that have a different slot for each day and time of the day (see Figure 14.1). This means you can keep track of what doses you have taken, which makes it easier to stay on top of your pain. It is a space saver — instead of carrying round lots of different boxes or bottles of pills, you just fill your box up at the beginning of the week, and carry that around. It is also easier to see when you are running low.

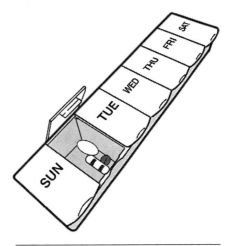

Figure 14.1: A dosset box is useful for organising your medication.

Pain Relief Tips and Tricks

◊ Leave some pain medication and a glass of water by the bed, in case you wake up in pain in the night. It's best to put the medication into a small pill box, for safety (out of the reach of children and pets).

◊ Take a decent dose of pain meds before you try to go to sleep.

◊ Have a checklist for your daily medication, and cross off when you've had your dose.

◊ Anticipate the pain (we know we've already mentioned this, but it's so important, we thought we'd say it again!).

◊ Don't run out. If you need more medication, organise this a few days *before* you run out.

Drug Dependence

Sometimes — especially if you're on some of the stronger pain medications — it is possible to find yourself physically or emotionally unable to manage without them. This is more likely to be a problem if you need these medications for a considerable period of time. This is not that uncommon, please don't feel that it's any psychological failing or weakness on your part!

If you are worried that you may become dependent on a medication, discuss it with your GP so they can monitor the situation. There are always ways to help with this, including changing the formulation of the drug (e.g. from short release to prolonged release), and tapering dose regimes to wean you off the medication. Stopping suddenly is rarely a good idea.

I had nerve damage after my THR, and was on Endone and Oxycontin for several months to deal with the neuropathic pain. I ended up getting physically addicted to these meds. It took quite a while to get off them, by gradually tapering down my daily dosage. The withdrawal effects when I tried to go "cold turkey" were really unpleasant, "crawling" skin, muscle spasms, insomnia, and so on. They also completely sapped me of any interest in anything, or drive to do anything much. **Denise, 47 Australia**

LOSS OF INDEPENDENCE

Loss of independence is something most of us have not experienced before major hip surgery, and hopefully won't experience again for a long time. It's surprisingly difficult to accept at first, especially when you're used to doing things your way. It's hard allowing someone take over for you.

Physically

If you are non-weight-bearing or on crutches for any period of time, your hands are your means of getting around. This means that you cannot use them for anything else (we know this sounds obvious, but you'll be surprised by how limiting it can be).

You can make a drink, but then you have to stand and drink it where you made it, as you can't carry it to a table or sofa (frustrating, hey?!) — the same goes for food, books, crafts, and the list goes on and on.

Around the House

There are some things you can do to help yourself (and the person looking after you).

Carry a little bag over your shoulder, on your back, or around your neck that can hold some basics (including a bottle of water or other drink), pain killers, glasses/spectacles, sweets, book, phone, pen, iPod and so on. There are instructions on how to make a knitted crutches bag in Chapter 19.

A small thermos is also a worthwhile investment. It allows you to carry a hot drink, especially if it has a good enough seal to be carried without leaking in your little bag.

Housework

Housework is a big no-no — the vacuum cleaner is not a walking aid! The twisting motions we do when sweeping or vacuuming are really tough on hips, and not allowed post-THR.

If you feel you cannot rely on your carer to clean or tidy to your standards, you either need to just let it all slide, or considering employing someone to do it for you, just while you are recovering. This has the benefit of reducing the stress on the people looking after you, and means you are not tempted to get up and do things yourself. Even if you are allowed to put weight through your hip, remember you have still had surgery, and it will take your body time to recover.

Cooking

Again, if you are non-weight-bearing, cooking is going to be difficult to do. As we have previously suggested, having prepared meals in the freezer makes things easier. If you're going to be on your own during the day, ask someone to make sandwiches or soup, for example, that can be kept in the freezer or fridge until you're ready to eat them.

If you have a family that needs feeding as well as yourself, you have more challenges. Forward planning is the best way to deal with this, and it's best if this situation has been anticipated before you go to hospital (see Chapters 9 and 11).

Our suggestions include:

◊ Teach your family to cook, even if it's just basic recipes. Have a list of easy recipes on the fridge, and the basic ingredients for them in stock ready to go (remember, a week of cheese on toast, while not perfectly nutritionally balanced, and rather boring, won't hurt). You can always "direct from the sidelines", too.

◊ Children can have meals at a friend's house occasionally.

◊ Change your diet for a few weeks, and just use prepared frozen or prepackaged meals from the supermarket.

◊ Ask some close friends if they can cook the occasional meal for you — this is a really concrete way for friends to help out, and most friends will be happy to oblige.

Psychological Impact

It can be very hard to cope with having a lot of your independence taken away from you, and however well you think you will manage, until you go through it, you won't know how it will affect you. Feeling useless, helpless, and frustrated may be some of the things that you encounter. It's worth remembering that your carer may also be experiencing these, too! This can sometimes lead to disagreements and arguments; but please keep in mind that this is something that you're going through together. Although it's hard to deal with, it is normal to feel like this, and it will get better. It is likely you will learn a whole new level of patience during this time! You can read more about these psychological issues in Chapters 7 and 9.

Social Impact

Honestly, at first it's unlikely that you will feel like going out very much at all after your hip surgery. This from a combination of recovering from the anaesthetics, side effects of painkillers, the damage done to your body by the surgery itself, and impact of dealing with pain and disability. This is entirely normal, and no reflection on your ability to cope. You will tire very easily; you need to take this into account, if you do decide to venture out.

Using a wheelchair will give you more flexibility in terms of being able to go out, however, even if you use a self-propelled wheelchair (highly recommended), it's unlikely you will be able to go out alone — another loss of independence!

DISABILTY AIDS

Disability aids are those nifty devices that make our lives so much easier and include all manner of long-handled tools (essential for when you're not allowed to bend over), toilet seats, cushions, and so on. We cover disability aids in full in Chapter 11.

Hopefully these will already be in place at your home, well before you left for hospital. Some long-handled grabbers come with a clip to attach them to crutches or walking sticks, so you can carry them around safely. It is a bit of a pain if you drop your only grabber ... what do you pick it up with? Having *two* grabbers to hand can solve this problem.

Keep the sock gutter nearby where you are going to be putting socks on. If you are struggling with the technique, practice on your "good" leg.

WASHING

This is one element of post-op life that may take some getting used. The level of help you need, and how exactly you bathe or shower, depends greatly on what facilities you have at home. Crutches and water don't go well together, and slipping is a very real possibility, so **do not wash alone**. In the early days at home, you will probably need someone helping you in the bathroom. Once you're able to wash yourself, always make sure there is someone in the house within earshot, who can help if you need it. We know of hippies who have slipped over while using crutches on a wet bathroom floor ... be careful.

273

It is unlikely that you chose your home with significant mobility problems in mind — it probably wasn't even a glimmer of a thought! So you may struggle to use the bath or shower, especially if your shower is in the bath, as you won't be able to step over the side of the bath. Strip washing may be your best bet for a while.

HIP TIP: Use Glad Press'n Seal® plastic wrap, which sticks to everything, including skin! Don't put a direct stream of water onto it, but it works well to cover the incision site while showering, and still have a dry wound area afterward. Much better than trying to tape plastic wrap over it! We're not sure if this is available in all countries, but it is definitely available in the USA. (Many modern surgical dressings are already waterproof, though, so it may not be necessary for you.)

How to do a strip wash:

◊ Strip off. Stand on a *secure* bathmat or a towel (nothing slippery).

◊ Fill the bathroom sink with (comfortably) hot water.

◊ Dampen your skin with a facewasher (flannel, washcloth), and apply soap or liquid cleanser.

◊ Working from the top of your body down, rub soap over your skin with the wet washer.

◊ Wipe off the soap, rinsing the face washer in the sink as you go.

Have a little stool (which is above your hips in height) that can go in the bathroom and the shower. If you haven't been able to get a medical one, plastic garden chairs can work, too (as long as they are high enough). Bath boards and non-slip mats are also aids that can mean you are able to get into the bath or shower safely.

Embarrassment

It can be quite hard to allow someone to help you with such a private daily task. Whether the person helping you is your partner, parent, friend, or an older child, it is likely that you will feel self-conscious (and they may too). It may take a few attempts to work out a method that suits you both. Try to keep a good sense of humour about the situation, a joke can help ease any tension. Trusting someone else to shave your legs is surprisingly frightening at first!

I was unable to shower as I could not step into the bath, so the alternative was to wash at the kitchen sink. This wasn't easy either, as I couldn't stand for very long. So my boyfriend found a solution — he fixed the garden hose to the kitchen sink hot tap.

In our back garden, there is one corner that is not overlooked by any neighbours, so as soon as the sun was out (which isn't often in London), I would hobble on my crutches into the back garden. I would get naked, and my boyfriend would quickly hose me down whilst I sat on a garden chair. If I was quick, I could just about wash myself and my hair before the sun went back in. During one sunny spell, I even managed to shave my legs! Funny ... haven't seen much of the neighbours since then! **Dani, 42 UK**

After my THR I went home from hospital to my flat that I shared with my boyfriend. I was laying in bed watching TV whilst he shaved my legs for me. I knew then it was true love! We got married last year. **Rhianna, 28 UK**

I used a plastic lawn chair instead of a shower chair (it had rubber on the bottom of the legs to prevent slipping, and armrests to hold onto while transferring). It hurt to sit on hard surfaces, so I used a newborn baby sponge tub as a seat cushion in the shower for the first couple of weeks. Showers were *exhausting* for the first few weeks, and I often took a nap right after.
Lara, 40 USA

DRESSING

One word here — **comfort**. Choosing clothes that are both easy to put on and comfortable to wear makes things easier during a time where getting comfortable can be difficult.

Avoid tight waistbands and clothes made of firm fabrics like denim. It is even worth buying and wearing some slouchy clothes a size or two bigger than your normal size (try second-hand shops for cheaper options).

Obviously, exactly what you wear will be dictated predominantly on what time of year you have your surgery. If you're going to be getting dressed by yourself, ask someone lay out your clothes the night before (within grabber reach) of the bed or a chair where you can sit to get dressed.

MOBILITY

Wheelchairs

Whilst a wheelchair can give you greater flexibility when going out in the early periods of recovery, you can feel quite self-conscious in one. As a rule, adults try to avoid looking at you, whereas children will stare — but practically no-one will ask you why you need it. We suspect that people assume that an elderly person in a wheelchair is understandable, but when people see a younger person in one then they assume something truly awful is wrong. Actually ... hip dysplasia probably qualifies on that count!

On the plus side, not having to explain yourself continuously to people is sometimes a relief, and customer service usually improves in shops, restaurants, and other venues.

It takes more planning to go out with a wheelchair or on crutches. It is worth checking both the methods of transport you are intending to use, and the places you intend to go, for disabled access. Make sure you check on such things as the presence of stairs, escalators, lifts, disabled toilets, and disabled parking places — do not assume that your needs will be catered for.

Crutches and Walking Sticks

As well as being an aid for helping you get around, these are a powerful visual cue for others saying "Do not bump or jostle me" and "Let me sit down." A backpack-style handbag or rucksack are essential for carrying anything you wish to take with you, or if you have shopping to take home. However, be warned — people are far more eager to ask you what's wrong when you are on crutches, which is nice in a way, but can get rather tiring after a while.

I was on crutches for two months over winter. Almost everywhere I went people were asking me if I had been skiing, assuming as I was young, I had injured my knee or something. That was one of the things I was looking forward to about getting off crutches — not having to explain myself all the time!
Sophie, 29 UK

I think when you don't have an obviously broken leg/ankle with a big plaster or bandage, people often stare when you are on crutches, and wonder what is wrong. I only used a wheelchair once after my THR operation, and that was to get some freedom whilst in the hospital for an hour. I actually felt safer on my crutches — in the wheelchair people do not "see" you, and are oblivious. I felt very exposed and worried about people walking into me, whereas on crutches you can create some space around you to protect the hip. **Deirdre, 36 UK**

When I was on crutches, people asked how I'd twisted my ankle, or broke my leg, and were very surprised when I told them I had had a THR! If they were strangers in a shop, sometimes they would just walk off after I'd told them. I suppose because they didn't know what to say to me? **Rhianna, 28 UK**

Because it wasn't obvious to anyone why I was on crutches, I would get lots of people staring at me while I moved slowly along on one leg with the support of my crutches. I felt like I had a whole school of children staring at me with curiosity whenever I went to pick up my son from school!

I only used the wheelchair a couple of times. At about five weeks post-surgery, my mother volunteered to take me out for a bit of retail therapy at our local shopping centre. I had lots of people stare at me with curiosity, but I think the people who stared the most were the elderly! **Chrissie, 31 Australia**

If public transport is going to be your preferred route of transport you may find that you need to use crutches, or a stick, for slightly longer until your muscles and stability are good enough to cope with all the jostling.

Generally I found people very helpful. Tesco's supermarket, in particular, would get someone to help me push the trolley round (they preferred it if I called ahead). Trying to get on an escalator in the Underground going for my six week check up was interesting — rush hour in London. Well, there is not a lot any one can do to help except be there to catch you! I never had anyone being rude, kids stared, so I just stared back! LOL! I actually found people trying to be overly helpful — which can be kind of irritating (in the nicest possible way) when you are trying to be as independent as possible! **Annick, 47 UK**

Physical recovery

The length of your recovery depends on what operation you've had. Other factors include: whether you've had complications, the state of your muscles pre-op, your general health, and the existence of pre-existing conditions like diabetes or auto-immune diseases. These all have implications as to how quickly you get over your surgery.

You need to be guided by your surgeon as to how long before they expect you to get back to normal life, sport, studies and work, as each case is different. Surgeons vary greatly in the restrictions they set post-operatively, so it is important to follow their instructions — the restrictions are there for a reason! The following are basic guidelines for the major hip dysplasia surgeries:

POST–ARTHROSCOPY

Most people are usually back to normal activities within a few weeks, and returning to sports and physical exercise at three months. If you had another procedure (such as a labral repair) done during the arthroscopy, your surgeon may want you to be non-weight-bearing for a period of time. If this is the case, it may be four to six months before you are back to regular exercise.

In the case of DDH, an arthroscopy is not going to correct the underlying structural abnormality, so your activities may still be limited after the surgery. An arthroscopy involves small incisions and minimal blood loss, and is not considered to be a major operation, however it is still surgery, and will therefore be a couple of weeks before you feel back to your normal self.

POST–PAO

The length of period you are non-weight-bearing affects your physical recovery process. Most surgeons require you to be non-weight-bearing for five to six weeks, followed by gradually building up to full weight-bearing status over the subsequent weeks. Most people are back to normal activities within three to four months (yes, it is a long old haul), returning to impact activites and sports after six months.

278

POST-THR

In most cases you are full weight-bearing from day two or three post-op, which means the recovery with respect to pain and activity is usually quicker than after a PAO. If you've had a cementless THR, you may need to be partial or toe-touch weight-bearing for some weeks, until the prosthesis is stable within your femur. A revision THR is a bigger procedure than the first one and therefore take longer to get over. However, you have the benefit of knowing what you are letting yourself in for (having already had a hip replacement previously).

Both PAOs and THRs are considered major surgery, involving blood loss, significant anaesthesia, and several nights in hospital, so it will take your body a fair while to get over the surgery. It can take up to sixs months, or even a year, before you feel back to your normal self.

OTHER PROCEDURES

There are too many variations and additional surgical procedures that you might have had on your hip for us to cover everything here in detail. The general guidelines for recovery from surgery apply — follow the advice of your surgeon and GP.

ANAEMIA

Most major surgery involves some blood loss. A blood transfusion, although safe with modern day guidelines and training, is not a totally risk free procedure, so most surgeons will only transfuse patients if the levels of haemoglobin in their blood fall below a certain level.

Haemoglobin is the iron-containing molecule in red blood cells, that carries oxygen around your body. Even if you have received a transfusion, your haemoglobin levels are still likely to be low, compared to before your surgery.

This means for several weeks in the early stages of recovery you may be anaemic. Fatigue, low energy levels, feeling weak and sometimes dizziness are all symptoms of **anaemia**. A good diet is usually enough to return your levels to normal, however iron tablets can sometimes be useful. Beware of side effects though! Iron supplements are notorious for causing nausea and constipation — it's worth trying several until you find one you can tolerate.

279

FATIGUE

Keep in mind that feeling tired and weak with low energy is all part and parcel of getting over major surgery, even if you are not anaemic. This is due to the operation and the stress it puts on your body, the drugs used during analgesia and pain medication, and so on. These things get better with time.

Sometimes it takes longer than you expect. This can sometimes happen because as you feel better, you start to do more, and therefore get tired more easily. If you suffer from other medical conditions, your recovery may take that bit longer than for someone who, aside from their DDH, is fit and well. If you are anything like us — the recovery will never ever be fast enough, even if you are doing very well!

CHILDREN

If you have children at home, their age greatly affects how much they understand, and can be involved with your recovery.

Very young children may find it hard to understand what has happened, and why you can't do all the things you normally do for them. Work out ways that you can still play with them and give them cuddles, without causing too much discomfort for you. While they may be able to grasp that physically there is something wrong (showing them scars or dressings can help), they are not likely to understand when you get exhausted quickly. Don't allow them to throw balls or other toys around near you.

Older children can understand more, and even help you out by bringing you things, retrieving stuff you drop on the floor, or carrying a plate of food or a drink out to you.

Older kids and teens can also help with chores, cooking and shopping. When you're stuck on crutches, it's impossible to shop without another pair of hands to get things off the shelves! It is important to thank them for their help, and not be critical. Most children enjoy the opportunity to look after you, and rise to the challenge of this role reversal, just so long as you don't expect too much too often.

My 2 year old daughter threw a six pound gel-filled stuffed animal up in the air a few days after I was discharged home. I was lying on the sofa. We all watched in horror, unable to react in time, as it flew up, then came down, and landed on my post-op hip! I screamed in pain and then started crying along with my daughter. We laughed about it later, but it was probably my most painful PAO moment. **Lara, 40 USA**

When my nephew (5) first saw me on crutches, I tried to explain to him that I broke my hip, and he said "I don't see it!", and I had to stop him from lifting up my skirt as he was looking for a cast or an injury! **Melanie, 29 USA**

When I launched myself through the front door on return from hospital after my first PAO, I nearly fell, and accidentally hopped twice on the operated leg! We all laughed later when we knew I hadn't done any damage, but I was caught by my darling son who (although I wish to think of him as a six year old and smaller than me) was 17, and much taller than me, thank god — otherwise he could not have caught me. **Annick, 47 UK**

The day I arrived home after the week in hospital, my son (5) and daughter (3) were very keen to help me out in any way they could, sweetly fussing around me. After a little while, I moved into my bedroom to have a lie down, but my children were concerned that they wouldn't be able to hear me if I needed them to help me with something when I woke up. So my son handed me a toy harmonica, and my daughter gave me a little cow bell, and I was to blow and ring until they came running in! **Chrissie, 31 Australia**

I have a 6 year old daughter, who was very upset about my surgery because I couldn't play with her as much, or she couldn't come to my bed at night. She was always asking me "Is this the leg that hurts?", and was afraid to hurt me more. During the first few days after I came home from the hospital, she tried to be an exclusive helper.

I also had my mom staying with me for some time, and my daughter didn't let my mom do things when I asked to bring something. She would even get angry, saying "It was me who she asked!", and trying very hard to be helpful. Although she admitted that my mom is allowed to help me with "disgusting stuff" — that being going to the bathroom and so on! **Arpine, 28 USA**

My three children (4, 7, and 10 at the time) would spend time with me, help wheel me around in the wheelchair, get me things I couldn't get, and even took turns putting on my socks for me! **Alison, 38 USA**

DVT STOCKINGS AND BLOOD THINNERS

If you've had any hint of a blood clot after your DDH surgery, you may well come home with those fetching tight and hot DVT (deep vein thrombosis) stockings — also called TED (thrombo embolic deterrent) compression stockings. You may also have to give yourself blood-thinning injections for a while; in some hospitals, this is a standard treatment after hip surgery, regardless of blood clots actually showing up or not.

Compression stockings are a pain to put on, no two ways about it, even on your good leg! You may need to use an aid for getting socks on (although these may not be robust enough for the compression stockings), or need to have someone to help you. There are some few tips and tricks in Chapter 11.

If you have to give yourself injections

Many blood thinners, like Clexane, are delivered by injection into the tummy, into the fat layer. You will be trained before you leave hospital.

This can be a real issue for some people, especially if you are needle phobic. The options for you are:

◊ You give them to yourself.

◊ Your partner or family member gives them to you.

◊ A district/home nurse visits every day to give them to you.

Obviously it is easier and more flexible if you don't have to rely on someone visiting daily to give you the injections, but we (and the hospitals) appreciate that it's something that some people just cannot do themselves, or to someone they love. The injections themselves are not that bad, the needle is extremely fine. They sting a bit, but compared to post-op pain, it's pretty minimal.

I hope not to develop DVT after my recent surgery; I am wearing the TED stockings constantly, and I am having Clexane injections into my stomach every night, so I really hope to not get a DVT. I am also moving around, which helps to prevent it too. **Sian, 29 UK**

I had a small pulmonary embolism after my THR, and had to wear DVT stockings for a few months (in a scorching Australian summer!), as well as take warfarin for six months. **Denise, 47 Australia**

LEG LENGTH DISCREPANCY

This is usually a factor for those people undergoing a hip replacement (and some types of femoral osteotomy) for DDH. Due to the nature of dysplasia and the degenerative process that leads up to needing a THR (and sometimes previous surgeries), some people have been walking round with their "bad" leg significantly shorter than the other. Long term, this can lead to strain on other joints and back pain, as you may well already know. During THR surgery, the surgeon *may* be able to correct any leg length discrepancy. However, the stability of the joint will always be their top priority; correcting leg length is always a secondary consideration.

Long term, adjusting leg lengths to be as equal as possible is better for your overall biomechanics. We say long term because, in the short term, it can cause more problems. As well as getting over the surgery, your back, pelvis, knees, and ankles need to learn to be in their "new" position. This means muscles will start to be used in ways that they have never been used before.

It can take up to a year for everything to settle down and for you to stop noticing the change. Even a such seemingly small adjustment as 1 cm (less than ½") will have a significant effect on your body. Nerves and tendons also have to stretch to accommodate the new leg length, and this is a very slow and often painful process.

As my leg has been lengthened because of my deep dysplasia, I was two centimetres taller on that side after my surgery, as a result of a bone graft. So until I had my second PAO, I had to wear special insoles in my left shoe.

I am just off my second PAO, and incredibly they have paired my legs. Now I am evenly levelled. **Alvaro, 38 Spain**

My "bad" left leg was nearly 2 cm shorter than my right leg, and had been rotated 30° outwards during childhood surgery. I've worn orthotics and lifts in my shoes for years. During my THR, 40 years after my last hip operation, my surgeon straightened and lengthened my left leg. For the first time in my life, my legs were the same length!

In the year after surgery I had sciatica, neural tension, knee pain, iliopsoas bursitis, a torn calf muscle (after kneeling down and getting up awkwardly), **coxodynia**, and a host of other minor but painful ailments at various times — all to do with my pelvis, spine, and legs getting used to the new state of affairs. My surgeon and physiotherapist said it would take about a year for my body to adjust completely, and it did. I am walking really well now, without any limp, and my posture is so much better, so in the long run has been worth it. **Denise, 47 Australia**

PARTNERS

If you are lucky enough to be in a relationship with a supportive partner, then this whole hip journey will be something that you experience together, and is hopefully ultimately, a positive experience.

Your partner will be taking over the role of doing everything that you would normally share, which can be difficult for you both — childcare, household chores, shopping, extended family duties, and so on. Tensions can get high, usually about little things that normally wouldn't bother you.

If you find this is happening, it is important to appreciate the other person, take a deep breath, and talk to each other openly and honestly. It's a good idea for your partner to get some time out by themselves. We cover this in more detail in Chapters 9 and 18.

PETS

If you have pets, their care is something you need to delegate to someone else. If you're on your own, you need to plan for this in advance. You may need help, for example, with walking a large dog daily, or feeding pets whose food and water bowls are on the floor. Many pets will pick up that you're unwell, and stay close, which can be a great comfort when you're feeling miserable.

If you have animals that like to cuddle, you obviously cannot tell them that you have a sore hip, and that they need to avoid jumping on it. Use cushions to protect that side, or sit in a way so they can still get some attention from you without hurting you accidentally.

Be conscious of hygiene around your pet and your healing incision — always use antibiotic hand wash, or wash your hands with soap, after playing with your pet.

I had a 6 month old puppy when I went for my THR, and on getting home from five days in hospital, it was great to get a lovely welcome. Due to the position of my scar it was OK for him to jump up on my knees when sitting, so we managed fine with that. For walks, I had my lovely mum help for a couple of weeks. By two weeks post-op, I was going with her on the walks, and walking my pup again myself by four weeks post-op. I wouldn't have wanted to walk him when I was still on two crutches. By the time I went down to one crutch, I felt confident enough to walk him by myself. The only challenge I had was picking up after him (if he did a poo), as after a THR you're not supposed to bend past 90 degrees. So I made sure we went places which were off the beaten track! **Deirdre, 36 UK**

My wife told me my two cats missed me a lot when I was in hospital because at night they meowed. **Alvaro, 38 Spain**

We have pets galore — as a dog sitter I had quite a busy summer, but also we have our own dog, two cats, two rabbits and two bearded dragons — the kids and my husband walked the dogs, I fed the dogs and cats, and the kids did the rabbits and lizards. Team work! **Annick, 47 UK**

My chihuahua Petal was a fantastic companion when I got home from my THR. She was rather worried about me, I think, and stayed close. She slept next to or on top of me as much as possible. Nothing like puppy cuddles and licks from Dr Petal to help one's recovery! **Denise, 47 Australia**

SLEEPING

This is tough at first, every position is uncomfortable! You may have to sleep on your back, with a pillow between your legs, for quite a while, which many people find difficult. But if you can improve your sleep, you will see a big difference in how you feel.

Sleep Tips and Tricks

◊ Take a good dose of **pain killers** just before trying to sleep. Keep some pain killers (and a drink of water) next to the bed so if you do wake up in pain they are within easy reach.

◊ **Pillows**: These can be used to prop you up, prop your leg up, allow you to sleep on your side with one between your legs, and stop you rolling round in your sleep. Work out how to make them work for you.

◊ **Sleeping tablets**: These can help you get a better night's sleep, however it is important to not rely on them as they can be addictive. Some sleeping tablets are also muscle relaxants, which can be useful if muscle spasms are keeping you awake. See your GP for a prescription.

◊ **Relaxation techniques**: Ask someone to massage your back and shoulders, or have a warm bath about 40 minutes before you sleep, or curl up with a hot water bottle.

◊ **Listening**: Listening to peaceful music, an audio book, or a CD of nature sounds like ocean waves or birdsong, can help. Some people find white noise the most relaxing.

◊ **Relaxation scripts**: There are many "sleep relaxation" CDs on the market, where the narrator guides you through various steps to get to deep relaxation.

◊ **Warm milk** is a well-known, and effective, drink to have before bed.

If you really can't sleep, there are a couple of things you can do:

◊ Get up, and read or watch TV. Many sleep specialists says that it's best to get up out of bed if you're not sleeping, because you want to set up a pattern of being "asleep in bed", and not "awake in bed". Only return to bed when you're feeling sleepy again.

◊ If you're drowsy enough to stay in bed, and don't want to get up, try not to stress about being awake, you're still resting and that's valuable. Listen to music or an audio book. Do some puzzles like words searches, or a sudoku, or cryptic crosswords, or play games on your phone. Getting all uptight and upset and worried about it only makes it worse.

Make sure you have a comfortable bed, and sleepwear. You may find using a softer mattress than usual is more comfortable. Your choice of sleepwear can even make it easier to move around in bed:

I have difficulty moving around and turning over in bed, and I have found that wearing pyjamas is a great help. When I need help to move, I just take hold of the PJ trouser waistband and use it to pull myself over. Sometimes it's the simplest solutions that are the best. **Margaret, 60 UK**

WOUND CARE

Surgical incisions need to be looked after carefully, to allow the wound to heal and a good scar to form. The precise material your surgeon has used to close the wound dictates what care you have to provide it. However, the following guidelines are true for most wounds.

You need to keep the wound completely clean and dry for 10 days. After this time, this the skin will have knitted together nicely, but will not be strong.

It takes two to four weeks for the wound to be strong, and up to a year for a mature scar to develop and become white. After 10 days it is okay to get the wound wet.

We do not recommend soaking in water, unless it is covered by a waterproof dressing. Showers, strip washing, or sponge baths are better than baths during this time.

Infection

Infection slows down healing, so if you have had a wound infection you may need to keep it clean and dry for longer. Signs of a superficial wound infection include: increased pain and redness around the incision, opening of the wound, leakage of pus, a slight temperature. You need to see your GP if there is any hint of an infection. You will probably need antibiotics. We cover post-op infections in more detail in Chapter 4.

Clips

Clips are like staples which hold the edges of the wound together. They only hold the skin together. The deeper tissues are held together with dissolvable sutures that you cannot see.

Stitches

Stitches are also called sutures. These may be dissolvable, or the type that need to be removed.

As a rule, clips and sutures are usually removed after 10 days to two weeks, depending on the wound. Dissolvable sutures that you cannot see do not need to be removed.

Glue

Some surgeons use a special type of skin glue which acts to seal the wound and can also make it waterproof. The benefit of this is that — while you still need a waterproof dressing as an extra barrier — it means that you can get the wound wet, which makes showering and washing easier, and also means you can start hydrotherapy almost straight away.

Scar Treatments

Once the wound has healed enough (after two to three weeks) and all clips/sutures (if present) have been removed, then massaging a plain moisteriser cream into the scar can help it mature. This massaging also helps to prevent it sticking to tissues underneath (tethering), which can make it look unsightly and become uncomfortable.

The emotional roller coaster

Everyone is different in how much they share their emotions with others. However you deal with them, we guarantee that there will be ups and downs in the recovery process.

CRYING

Let's face it, everyone cries. We all know there are times that welling up just can't be stopped. This is perfectly okay, and it's fine to do it a lot too! There are lots of reasons why you will feel more emotional than usual in the first weeks post-op: drugs, anaesthetics, pain, lack of sleep, frustration, and vulnerability, to name a few. So if you find yourself crying at the slightest thing, please don't worry, it's quite normal after such major surgery.

My children sent pictures to me at the hospital. When I came back to home I started crying. **Alvaro, 38 Spain**

I think I cried every day after my THR, for about two months. Pain, frustration, crushing fatigue, anything really — things I could normallly handle well were all too much for me. **Denise, 47 Australia**

I was very emotional when I got home from hospital. In fact I think it started in the car on the way home! I was totally unprepared for this. Even looking back now I still don't really understand it. I know I was hugely pleased to be home. as I had been in hospital for almost two weeks. The operation had gone fine, but there was trouble sorting out my pain relief.

I guess my old words of advice would be not to worry if you do feel down for a few days. It's all part of the healing process and it will soon pass. **Dani, 42 UK**

I was in hospital for a week, and thought I was feeling okay, being strong and sensible about it all, but the moment I got home, I just burst out crying! It came as a real surprise — it just felt like all the fear, tension and quite a bit of feeling sorry for myself just couldn't be held in any more. I just sobbed and sobbed — it was a bit scary — I felt very out of control.

It took a long time for the fear to go away — fear of falling over, of hurting myself, or dislocating my hip — fear at every small pain or twinge. The fear became tiring and made me sad — was it always going to be like this? But slowly, as my hip healed and became stronger, a feeling of growing happiness took over — it had worked! It took a while to feel this good. Surgery puts you through a lot, both physically and emotionally — be kind to yourself, not impatient — give yourself time to heal. **Freja, 45 UK**

We hope that you find the following "recovery" experiences heartening; while many of these hippies had a rough time of it initially post-op, you can see that they've all come through it, and that things do get better!

I was pretty emotional in the first two months after surgery. I began to despair that this constant pain, and limited mobility was as good as I would ever be. I found times when I wondered if I'd made a mistake doing the surgery, and feared that I might get addicted to pain pills. The evening injection of blood thinner into my abdomen became a dreaded event. The pain wasn't that bad, but just having to have it done was bad, and my husband hated hurting me, and I hated knowing that he had to do this. I felt less of a person for needing so much help to do so many things I'd always been able to do. I felt less of a wife to my husband, because all of the family chores and caring for me fell to him.

I was fortunate to have very supportive husband, mother, and friends. They brought me chocolate brownies, and rice crispy treats, which are two of my favorites, and lifted my spirits. I also read the HipWomen stories, and found strength and patience from them, it helped me knowing that what I was going through was normal and certainly not permanent. When my emotions were darkest, I watched TV shows that were uplifting or humourous, and also made note of the little things I was able to do, when I could roll over with less discomfort, when I could increase the range on the CPM (continuous passive motion) machine, when I went longer and longer between pain pills. Little stuff like that became super important to improving my emotional state.
Lisa, 40 USA

290

Bri (15) was very tired! Exhausted from anemia. Frustrated at having others have to help her. But this period was very temporary. Bri was really touched by the kindness shown to her by other people. She had spent months in pain with no outward sign of what was happening to her, so people didn't really reach out to her. When she came home, people had learned what she had been through and let her know how brave they thought she was. **Cynthia, 47 USA**

The first few weeks were okay, although I got increasingly frustrated in being in a recliner most of the day. I wasn't in nearly as much pain as when I had ACL (anterior cruciate ligament) reconstruction, but it was a different kind of pain and discomfort. To keep your sanity, I would recommend a strong support system. I was in touch with friends everyday, had occasional visits when I was up to it, and my family was at my house every day.

I was lucky to have a doctor who was in touch via email, and encouraged me to check in every week. That was a comfort, because I was able to tell him everything that was going on and what I was feeling, and he would get right back to me. My doctor was positively the best, and without his support I'm not sure I would have had the same experience. He's unlike any doctor I've ever been to, which is why I chose him after seeing five surgeons for my hip issues. I don't know any other doctors who keep in touch on a weekly basis with their patients via email or by phone. He was amazing. **Alison, 38 USA**

Overall the experience was pretty traumatizing. Some nights I have nightmares of being in the operating room with the bright lights over me. I do not want to go through this again, although the cruel reality is that I'm going to have to, sooner or later. **Juan, 34 USA**

After my first PAO (left hip) I was absolutely fine emotionally, and couldn't wait for the next op! **Annick, 47 UK**

I came back from my THR with obturator nerve damage, permanent hip restrictions, and a pulmonary embolism. For quite a while I felt that I had just swapped one disability for a bunch of new ones, and was quite depressed. I'm over that now, though! The complications have healed, and I've adjusted to the permanent restrictions. **Denise, 47 Australia**

For the first week or so after I got home, the meds left me pretty out of it. However I was more emotional than usual; fits of crying, some "down" days. I just accepted that it was part of the initial recovery phase after a major surgery and pushed through it; lack of sleep was also a *big* factor in the depression and sadness. By the third week post-op, my spirits started to lift, as I stopped taking the narcotic pain-killers and was able to get out of the house with friends. I was starting to sleep better at that point as well — it is amazing what a few great nights of sleep can do for your mood! **Kris, 47 USA**

Emotionally I was fine after the expected day three dip post-op. I struggled with sickness, and had no appetite for a good four or five days post-op, so that made me feel a bit miserable. I soon perked up, though, and was just delighted to have my new hip, and that carried me through the restrictions, and being stuck at home. I was also able to do some work at home (self employed) at two weeks post-op, so that also helped to pass the time. **Deirdre, 36 UK**

I felt a loss, and experienced all the phases of loss/death. But we are also initiated into a more bionic life ... So it's a tradeoff I suppose! **Jodi, 56 USA**

I was very pleased to sleep at will, and read in quietness, and be spoiled by my wife and daughters. The worst part were the crutches and having to sleep on my back. **Alvaro, 38 Spain**

It's quite difficult when you get home because you suddenly realise just how much you're *not* able to do. Even the easiest of tasks like getting in and out of bed, showering, and putting on underwear become a struggle. And it's quite disheartening to be so dependant on others. After a PAO, especially, the first two weeks are the toughest because you don't really see any improvement. But once you start to see some, you know that although it may be frustratingly slow, you *are* actually recovering.

COPING STRATEGIES

Everyone copes in different ways. For most people, undergoing major surgery for hip dysplasia has been one of the biggest decisions of their life. The thing with this type of surgery is it is not a life-saving procedure, but is all about quality of life.

It may help in the low moments (and there will be some) to think of all the things that you can look forward to doing, even if this may not be possible until after yet more surgery (if you have bilateral dysplasia).

Goal setting is another strategy that can help, but please remember not to set yourself unrealistic goals, as failing to achieve them will make you feel worse. For example, the goal of "I will do five minutes on the exercise bike" is more realistic than "I will go on a two mile hike this afternoon". Reward yourself for the steps and achievements you make, and it is perfectly fine if they are only baby steps: the first time you clean teeth standing up, your first hydrotherapy session, your first solo bath/shower, and so on.

Acknowledge the small milestones, look back and take notice of what you have achieved so far. It makes the long road ahead appear easier.
Sandra, 39 Australia

Try and find something that will distract from sad thoughts, and help you feel accomplished. Like a craft, or small projects, writing — something where you can do things and see the end result. **Arpine, 28 USA**

I was a wreck emotionally the first few weeks after my surgery. Frantic about not being in control of my life, being dependent on others, not being able to take care of my kids, feeling sick and tired and helpless. It was harder to deal with that than the physical pain of PAO. I was prepared for the physical pain, but not so well emotionally prepared.

Although I was not able to do this, I recommend allowing yourself to give up the control for a while, accept help from others, take one day at a time, and most importantly — rest, and let your body *heal*. **Lara, 40 USA**

My advice, at least from a THR perspective, is to recognize you've had major surgery. Your moods will be up and down as a result of the operation, the medication and your sleep patterns. It *will* pass eventually — hang in there with it as best you can. Get outside as much as you physically can at first. Even walking around my own driveway was heaven. My husband took me for a lovely country ride about 10 days after surgery, and that was such a gift. Planning trips with my friends also gave me something to look forward to, on days when I was alone all day long with no way to get outside. **Kris, 47 USA**

SETBACKS

Hopefully you won't have any of these, but it's as well to be prepared just in case. Setbacks can include things such as: developing complications from your surgery, the other hip causing you more problems, having an important medical appointment postponed or rescheduled, your surgeon moving away, getting an infection, or your weight-bearing status not changing when you hoped it would be.

The key here is to take things one day at a time, and try not to catastrophise. Remember why you're on this hip journey — and we know it sounds trite, but there really are people worse off. You will get over it, and things will change and hopefully improve, even if it isn't the path you anticipated being on.

After my second PAO, I was a basket case! The undiagnosed fracture caused me *huge* amounts of pain, and constant discomfort from clicking and moving about even when I was stationery. I buried myself deep inside myself (yeah, depressed!) and didn't want to see anybody, talk to anyone or do anything. It took me six months to really feel back to my normal self. I had to cope with it on my own, as nobody close to me really understood. All they kept saying was "Oh well, the second operation is always different to the first, so you can't expect it to be as easy a ride as the first one was". Nobody really tried to read between the lines and figure out that something was wrong — that was left for me to do.

It was not a nice time and I would say to anyone: You know your own body — if you feel there is something wrong, pursue it, and keep on at people until you get to the answer. It took me three and a half months, but I finally had my concerns that something had been wrong confirmed, when my stress fracture showed up on the x-ray. I *knew* I hadn't been 100% for a reason, and I was right. **Annick, 47 UK**

RECOGNISING WHEN YOU NEED MORE HELP

We've said that the whole recovery process is an emotional roller coaster, however it's important to recognise in yourself that overall these things should generally be getting better a little bit, day by day, even if it seems slow.

If you find that things are not really improving (emotionally, not physically, although one may be linked to the other), talk to your GP. It is not unusual for major events in life (this is definitely one of those!) to throw you into clinical depression or a persistent low mood.

There are several options available to treat this, from counselling and cognitive behavioural therapy (a special type of counselling that is particularly good for coping strategies), to medications, so don't suffer in silence. Needing medication for depression or anxiety is no failing on your part, too, it can be remarkably effective, even if it takes a while to find the right medication.

Here are some of the symptoms of depression to look out for:

◊ Low mood.

◊ An inability to experience pleasure (*anhedonia*).

◊ Loss of interest in things you usually enjoy.

◊ Loss of appetite or increased appetite.

◊ Weight loss or gain.

◊ Sleep problems, and early morning wakening.

We discuss depression in more detail in Chapter 7.

We hope this chapter has given you some ideas of what to expect when you first come home from hospital, and some strategies for helping make this time a bit easier for you. And congratulations, you're well on the way to your new life with great hips!

295

Chapter 15
Living with Your New Hip

After you've had your hip surgery, especially a total hip replacement or revision, there are likely to be a lot of things you need to do and remember to keep your new hip happy. Some of these are short term while the damage to your joint repairs, and others are long term. There are some things you need to keep an eye out for as well, to stay as healthy as possible with your new hip. Infection and **aseptic** loosening are the major bugbears.

Hip restrictions, or precautions, are the set of guidelines that your surgeon will give you after your surgery. They are most strict after a total or revision hip replacement, but you may still have some restrictions after a PAO or a resurfacing. Generally, hip restrictions are short term. However, if you can maintain a low level of hip restrictions permanently, it will help the longevity of your prosthesis.

THR hip restrictions

If you have a total hip replacement, you will have hip restrictions after your surgery, generally lasting for three months. These are limitations on how you're allowed to move your hip and leg, and help to guard against dislocation. If you have a cementless prosthesis, you may be partial or toe-touch weight-bearing on crutches for six weeks or so. With a cemented prosthesis, you will be encouraged to be full weight-bearing as soon as possible after surgery, usually within a day or two.

The joint capsule, which holds the joint securely in place, has been cut up, damaged, and stitched back together — so it is very weak for quite some time. The three month period is what your body needs to have those muscles, tendons, and ligaments restabilise and repair fully.

After THR surgery, you need to observe certain precautions for the first six weeks to prevent the ball from popping out of the socket (dislocation). Your surgeon and physiotherapist will provide you with exact information and the time you need to adhere to the restrictions.

297

BASIC HIP RESTRICTIONS

◊ Do not bend your hip beyond 90° (a right angle), bringing it close towards your body.

◊ Do not cross your legs.

◊ Do not let your operated leg move past the midline of your body.

◊ Do not twist your operated leg.

◊ Do not bend over. Use a grabber to pick objects off the floor.

◊ Do not sleep on your side or stomach until allowed by your surgeon.

THE 90° RULE

Don't bend your operated hip beyond 90°. Some surgeons may set a lower angle, such as 80°. This one rule will probably have the biggest impact on you. Some things you mustn't do as a result:

◊ Don't lift your knee higher than your hip.

◊ Don't sit on low chairs, car seats, or sofas — you will probably need extra cushions on them (firm foam blocks work well). The seat must be your knee height or higher.

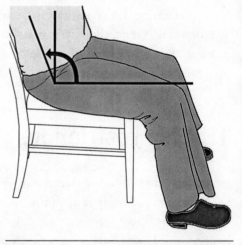

Figure 15.1 Correct position for sitting after THR. Maintain a 90° angle between the torso and legs, or wider.

◊ Your bottom should never be lower than your knees.

◊ Use an elevated toilet seat. This looks like a chair frame with a toilet seat in it, which sits over your own toilet. You can also get "clip on" toilet seats that attach directly to the toilet bowl.

◊ Don't lean forward when you're sitting down. If you're working at a desk, arrange your keyboard and chair so that you can sit straight in your chair. Use a wedge cushion on your chair.

AVOID THE MIDLINE

Your operated leg must not cross over the imaginary midline of your body. Don't allow your legs to cross. This means sleeping with a wedge or pillow between your legs when you lie down (even when flat on your back). Keep your legs 10–15 cm/4–6" apart when sitting.

Put a pillow between your legs when you're lying on your side (Figure 15.2). Do not sleep on your side until your surgeon gives you the all clear.

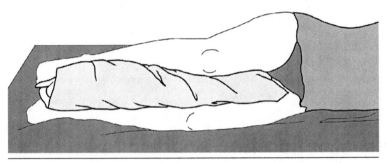

Figure 15.2 A pillow between the legs stops the operated leg (on top) from crossing the midline.

TWISTING AND TURNING

Depending on exactly how your hip replacement was done, you may have some or all of these restrictions, in addition to ones above:

◊ Don't turn your operated leg inwards. This is called **internal rotation**, and happens if you twist your body *towards* the operated hip, without moving your feet (i.e. "pigeon toed"). To turn around on the spot, shuffle your feet around. (Figure 15.3)

◊ Don't turn your operated leg outwards. This is called **external rotation**, and happens if you twist your body *away* from the operated hip, without moving your feet (i.e. "duck footed"). Again, to turn around on the spot, move your feet with your body, shuffling around on the spot. (Figure 15.3)

◊ Don't hyperextend your leg (i.e. kick your leg out to the back).

◊ Do not lift your leg with the knee straight, it puts too much stress on the hip joint.

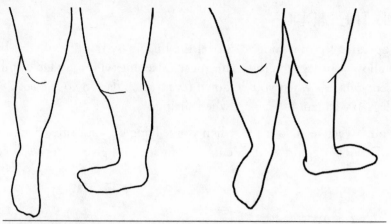

Figure 15.3 Internal rotation (pigeon-toed) on the left, external rotation (duck footed) on the left. In both cases, the operated leg is on the right.

Hip restriction tips and tricks

The disability aids mentioned in this chapter are described in more detail in Chapter 11.

BED

It's best if you can get in and out of bed on the same side as your operated hip — so if your *left* hip was replaced, sleep on the *left* side of your bed (as you are lying in it), and get in and out from the *left* side. Your operated hip should be on the edge of the bed. This also means if you are sharing a bed with someone, they are not going to accidentally bump against your sore hip! You need to sleep on your back, very possibly with a pillow between your legs, for around four to six weeks. Although this is one of the things that many hippies hate the most during recovery, trust us, you will be so exhausted so much of the time in those weeks, that you will manage to sleep regardless.

CARS

To get out of the car without stressing your hip, do the "princess move" — swivel around on the seat until you're facing the open door, and move both legs so they're out of the car, with knees together (Figure 15.4). Then stand up, either holding onto your crutches, or the car for support. Decorous, elegant, *and* "hipsafe"!

Don't do the usual "twist and turn" move on one leg that you're probably used to. Sitting on a plastic bag can help with swivelling around (you can buy a special swivelling pad, but they're fairly expensive).

To get into the car, again, sit on the side of the seat, then move your legs around so you're facing forwards. You might need to use your hands to move your operated leg, if the muscles are still sore and weak. Be careful, as ever, of the 90° rule, and not leaning forwards.

It's a good idea to set the back of your car seat to a wider angle, lying backwards more, so when you sit back your hip maintains a nice open angle. When you're a bit more recovered, keep your operated leg straight and out to the side when getting in or out of a car.

Getting in and out of a car post-op (or even pre-op) — I didn't find the circle-turning cushions any use and went far more technical: pop a heavy-duty plastic carrier bag on the car seat and sit and glide/turn into position. Obviously have the seat positioned, and reclined as far back as possible. It has to be heavy enough plastic that it doesn't scrunch up, i.e. one from "posher" shops. Avoid low cars, consider a wedge cushion to raise the seat, and pop it in the plastic carrier bag — one of these is a permanent feature in my car. **Margaret, 60 UK**

Figure 15.4 The princess move for getting out of a car after THR.

CLEANING AND GARDENING

Basically, **don't**. Get help. Ask your family to do it. Pay for a handyman or cleaner, just for a few times. You must not do *anything* with a twisting motion after a THR, so chores like vacuuming, sweeping, washing floors, and raking are out (Oh curses, we hear you say!).

Even if *you* feel up to the task, **your hip isn't**. You may need to accept lower standards with the housework, but just let it go until you're fully recovered. If friends visit and want to know what they can do to help — hand them the mop; you'll be surprised, they're likely to be very willing to do something so practical.

If you absolutely *must* do weeding, use a long-handled tool for this task, but avoid it if at all possible. You might like to invest in a Roomba robotic vacuum cleaner, we can recommend them from personal experience! They certainly make that task a lot easier. They're not cheap, but do work well, and take all the tedium out of that particular chore. Plus they're entertaining to watch!

COOKING

Have all your equipment at waist-height and within easy reach. Before you go in for surgery, move your daily crockery and cookware to shelves or benches that are waist-height and above (i.e. you don't want them in low-down cupboards). Also avoid having frequently used items in high cupboards that you have to use a foot stool or step ladder to get to. Your goal is to *not* have to get down low, or climb up high, for anything for a few months.

CRUTCHES

Use your crutches mainly for balance, post-THR you will be encouraged to weight-bear just about from day one (unless you have a cementless THR, in which case you may be only partially weight-bearing for some weeks). Hopefully you can get off the crutches and onto a walking stick after four to six weeks. When using crutches, try to remember to keep your chest up, shoulders down, and tighten your bum muscles and core/tummy muscles. Your physio or occupational therapist will help you with this. They will also teach you how to safely negotiate stairs before you leave hospital.

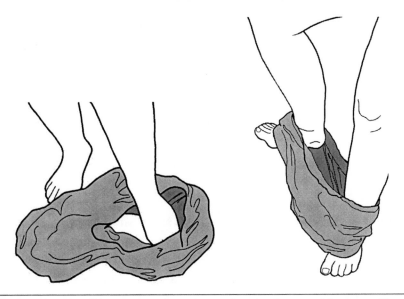

Figure 15.5 How to pull up your pants or a skirt without bending over. The operated leg is on the right in both illustrations.

DRESSING

Use long-handled aids such as a grabber, shoe horn, and sock gutters (see Chapter 11 for more details on these items). Have your clothing items within easy reach. Make yourself Freja's Dressing Clips (using clips and string) in Chapter 19. In the early weeks post-op, wear clothes that are easy to put on and take off. You may need help with things like underpants, socks, shoes, and trousers.

A "hipsafe" way to put pants, skirts, and undies on is illustrated above in Figure 15.5. First of all, drop your pants onto the floor, and arrange them with your foot or walking stick so the leg openings are clear. Step into the appropriate hole with your *operated* leg. Then use your *good* leg and foot to catch the other side, and lift up the clothing, with your foot as the hook. Then you can grab the edge of the fabric, and pull your pants up!

DRIVING

Each state and country has different laws regarding driving after THR surgery, so check with your local road transport authority or hospital. In some places you will not be covered by accident insurance if you drive within this "exclusion time".

It's not advisable to drive for quite a while, at least six weeks. If your new hip is the one that operates the brake and accelerator, you will probably need to wait longer. If you drive an automatic car, and the hip that was replaced isn't used in driving (so left hip in the UK and Australia, or right hip in the USA), you may be able to get behind the wheels a bit sooner. See Chapters 6 and 16 for more information on this.

PICKING THINGS UP

Let's face it, there's a limit to how many weeks you can keep calling on your family to pick up anything you drop. Mechanical grabbers — while generally very helpful — aren't always the right tool for the job (such as refilling a pet's water bowl). If you need to get something from low down or on the floor, you can try the "ballerina' move" (Figure 15.6). Hold onto something sturdy for safety. Stick your operated leg straight out behind you (i.e. without bending at the hip, and with the knee straight). Be *very* careful that you don't fall over, though; this move is best used sparingly at first. It is best when you're off crutches, are steady on your feet, and have reasonably good balance. ***Always*** hold onto a bench, table, or chair back for extra stability. One surgeon has said he advises all his THR patients to use this posture when picking things up, permanently. Once you're fully recovered, you don't need to hold onto something for support, if your balance is good.

Figure 15.6 The ballerina move for picking things up off the floor safely.

SEATING

You must not sit on low chairs for three months because of the 90° rule, so you need to find suitably high seating in your house. The seat should come to the *back of your knee or higher*. For sofas, you might like to buy a foam block to put under the sofa cushion, or put on top of other chairs. If you're feeling extra crafty, you can cover it with fabric to make it a bit more attractive and durable. You can also buy foam wedges (Figure 11.3) and blocks from disability aid suppliers. Foam wedges are good, as they help you maintain that open angle on your hip. Some people like to buy, borrow, or rent a special high seat chair, which are designed specifically for hip patients.

SHOWERING

Have all your toiletries within easy reach, on a shelf or shower caddy. Use long handled sponges and brushes. Soap on a rope can be good too, or liquid soap in a dispenser or bottle. If you're still a bit wobbly on your legs, use a shower stool to sit down. If you drop anything, *don't bend over* to pick it up! Just leave it. Avoid baths for the first six weeks, to avoid the wound and dressing getting too wet and icky (technical term there). If you have trouble stepping over the side of your bath, you may need to resort to sponge baths for a while. Chapter 14 goes into more detail.

SITTING DOWN

Back up to the chair or bed until you feel it against the back of your legs. Reach behind you for the bed or armrests of the chair, and slide your operated leg straight out in front of you. Remember not to lean forward as you sit! Try to always sit in a chair with arms.

STANDING UP

Push up from the bed or chair, and keep your operated leg straight out in front of you. Raise yourself without leaning forward. This movement is one of the trickier ones to do with hip restrictions.

TOILETING

Be careful of how far you lean forward when sitting on the toilet, remember the 90° rule. Make sure the toilet paper isn't behind you, so you don't have to twist around to get to it.

Going to the toilet when you're out of home can be very difficult, especially as many disabled toilets actually have *low* seats! Urination is obviously easier for men here, but women can use a device for peeing standing up (see Chapter 11). If you do need to sit down, you may need to develop strong thigh muscles so you can "hover" over the seat, and brace yourself on the walls of the cubicle. If visiting friends, you may be able to take your toilet seat raiser with you (especially if it's the "clip on" sort).

Incontinence pads are a good stop-gap measure, they can hold several cups of urine.

TRIPPING HAZARDS

Be careful of tripping hazards around the home, so tidy up any mats that might slide, and cords that might trip you up. Do anything possible to reduce the risk of tripping and falling. Also be wary of any pets who have a tendency to dash suddenly between your feet!

HIP TIP: This is a hip-safe stretch for your thigh muscles, if they're feeling tight and sore. Stand side on to a chair with your good side closest to the back of the chair, hold on to the back of the chair. Rest your operated leg's knee on the seat of the chair. Pull back on the foot with your hand. Hold for 30 seconds. Repeat if so desired.

Figure 15.7 A safe thigh stretch.

WORK AND STUDIES

If you go back to work or studies before the three months of hip restrictions are over, remember that you need to keep doing those restrictions at work too. Make sure you don't lean forward at your desk. Take in a cushion to raise the level of seats. You might like to get a letter from your surgeon, GP, or physiotherapist, explaining what adaptations you require, if your usual work or study activities have to be modified because of the restrictions. There is more about this topic in Chapter 9.

PAO hip restrictions

As after any surgery, avoid having baths and swimming for a good two to three weeks, or however long it takes until the wound is *completely* healed, and your stitches and dressings are off. Showers or sponge baths are the way to go. If your consultant uses a special glue called Dermabond™, this creates a waterproof barrier, and means that you can be less careful with regards to getting the area wet. This can allow you to start hydrotherapy straight away after surgery, even before you leave hospital in some centres!

The restrictions post-PAO can vary depending on your surgeon and exactly which muscles and tendons are disrupted during surgery:

◊ No straight leg raises (lifting leg off the bed without bending your knee) for six to eight weeks (Figure 15.8). This is because the muscles that do this action are usually cut (and then repaired) during the procedure, to allow access to the joint. If you do this too soon then the muscle or tendon can rupture or break. Not only would it hurt, but likely need to be surgically repaired again!

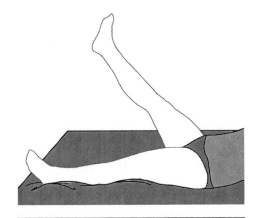

*Figure 15.8 Straight leg raise. This move is **not** permitted for several months after THR and PAO surgery.*

307

◊ No full weight-bearing for six to eight weeks. Although your bones have been fixed securely together with screws, there are still gaps in the bones that take time to heal. By six to eight weeks post-op, the bones are solid enough to hold your full body weight, though it may take up to six months for the bones to completely heal.

◊ Most surgeons keep you non-weight-bearing or toe-touch weight-bearing (just touching your toes to ground for balance) for the first six to eight weeks, before gradually progressing to full weight-bearing over the next few weeks.

Apart from these, there are remarkably few PAO restrictions, compared to a hip replacement. Pain is usually the limiting factor with how you move. There are no restrictions to moving your hip into **flexion** or rotating it, however you will probably find it is too sore to be doing much for a while. Be guided by your surgeon's post-operative instructions, and what your physiotherapist tells you (as long as the physiotherapist has experience with PAOs, otherwise just listen to your surgeon!)

I had no PAO restrictions, although I was told marathons are not recommended!
Sandra, 39 Australia

Resurfacing restrictions

In general, if you've had a resurfacing, you may be given the same hip restrictions as for a THR patient for six weeks (see above). Or you may have no restrictions at all! You may have partial weight-bearing, or a mix of weight-bearing and crutches.

Basically, there isn't a standard procedure with resurfacing, so follow your surgeon's advice. Your surgeon may order the standard hip restrictions as they have a proven track record in preventing dislocation in THRs, and they figure "Why not? It can't hurt to be careful for a little while."

Resurfacings have a very low rate of dislocation to start with, though, so if your surgeon doesn't give you restrictions, don't be too concerned. Resurfacing is a relatively new technique, compared to hip replacement, so the field and standard methods are constantly evolving and changing.

One major difference between a hip replacement and a resurfacing, however, is a higher risk of the femoral neck fracture with a resurfacing. If your surgeon gives you restrictions to guard against this in the weeks immediately following surgery, follow them to the letter! You may have limits on how much weight you're allowed to lift and carry for some months, for example.

Personally, I was happy to have an excuse to be very careful and cautious after my resurfacing. With a 10 inch incision, and 42 staples, I didn't want to break the 90° rule in a hurry, anyway! **Freja, 45 UK**

Currently, four months out from my resurfacing surgery, my doctor doesn't want me doing long-distance walking yet or jogging. I asked him if I could walk a 10K in a few months, but he felt it was too early for the pounding of walking that distance. Other than that, I can pretty much do anything. I do have to watch myself every now and then — especially twisting and turning. I feel a pinch now and then, but nothing that's a problem.

During the initial recovery period post-op, I had pretty much the same restrictions as a THR. Initially, I was pretty nervous about those restrictions, but once I had the surgery it was all pretty easy. The biggest ordeal was trying to sleep on my back — once I could flip over, I was a happy camper. **Laurie, 50 USA**

I had no hip restrictions after my resurfacing. I was encouraged to push the boundaries, and explore what range of movement I could achieve. I was walking with sticks for eight weeks, and then had to wean myself off them and focus on walking without a limp — which took time (I did have a lot of muscle wastage though). I'm sure recovery is quicker for some.

As I had waited so long to seek help for my dyplasia, my muscles were very weak, and I had to work hard in rehab. I spent a lot of time in the water walking and running before I could do weight training to improve my strength. It was tough, but very doable. Because I like to be active, I was OK with the process. The pain was minimal and in no way did I suffer like I did after my PAO.

I now feel a real need to look after my hip. I would love to run more, but restrict myself to more low-impact exercise. I fear having surgery again, so want to give myself the best chance of this working. **Tina G, 41 UK**

Permanent hip restrictions

In rare circumstances, your surgeon may tell you to maintain some level of hip restrictions permanently. If so, there is always a good reason. Your particular prosthesis may be slightly more prone to fractures, for example, or your particular anatomy and deformity may make dislocation more likely.

Generally, permanent hip restrictions involve things like not participating in high-impact sports, not running or jumping, not lifting heavy weights, and not doing certain stressful movements (like bending your leg at an extreme angle close to your chest, and twisting it around at the same time).

While it may seem like you're swapping your previous disability for a whole *new* set of disabilities, they won't be as tough as those early three months of restrictions, and you will get used to them very quickly to the point where you barely notice.

However, if you are used to playing a lot of sports, such lifelong restrictions could be pretty devastating. It's likely you will need to completely stop activities like gymnastics, aerobics, martial arts, running, basketball, soccer, netball, and so on. In this case, you may want to seek some support from a counsellor if it really impacts heavily on your lifestyle, and is very upsetting to you. It's worth exploring creative and positive ways you can adjust, perhaps finding other sports that you would also enjoy that are low-impact and safer for your hip (such as cycling, rowing, walking and swimming), or maybe moving into coaching, if you are a professional sportsperson.

My femurs are abnormally small (a deformity often seen with DDH), and my surgeon had to specially order in the smallest hip prosthesis available for my THR. Even then, it was a bit big, and he had some trouble getting it to fit! Unfortunately, this particular prosthesis does have a higher risk of the stem fracturing, so I have permanent hip restrictions to help avoid this happening. There are a few extreme twisting and bending movements I mustn't do, and I have to avoid high-impact activities, but they have little influence on my everyday life. I just have a great excuse not to join in any fun runs now! For me, the permanent hip restrictions are a good trade off for reducing the risk of a fracture and emergency surgery. **Denise, 47 Australia**

I have one permanent restriction after my THR, and that is not to put weight on the operated hip when the foot is turned slightly inwards, and swivel. It's a no-no for THRs as it can cause dislocation. And the no crossing the legs rule too.
Sandra, 39 Australia

Even though you may not have been given permanent hip restrictions, your hip replacement will last longer if you avoid high-impact activities, heavy lifting, jumping from heights, falls, and carrying heavy loads. It is always best to discuss your individual circumstances with your surgeon and physiotherapist. Gaining weight is a problem too, as it puts more stress on your hip joints. Try your best to lose weight if you're over your most comfortable weight (we know, easier said than done!). See Chapter 5 for help with this.

Some surgeons will advise you to follow some level of hip restrictions permanently after your surgery, to best care for your hip. Sitting on very low seats, doing squats, and twisting on your operated leg, for example, are all stressful to the joint.

Serious THR complications

In this section we discuss the main serious complications that can arise with a THR. This is in addition to the post-op complications mentioned in Chapter 4. The incidence of these complications is low, but there are some things that you need to be careful of, for a rosy future with your new artificial joint.

NOISY HIPS

Having a noisy, squeaking hip prosthesis is generally not a serious complication or a medical problem that requires treatment. It's more of an annoyance. However, in some rare instances, it can be a sign of loosening of the prosthesis. Because of this, it's important to inform your surgeon if your artificial hip gets noisy.

If the noise is driving you completely to distraction it can be treated — but the correction involves a revision surgery (see Chapter 4 for details). Basically, figure out which movements and positions cause the noise, and try to avoid them.

There are several reasons you might suddenly find yourself making more noise than you'd like (there go your ninja sneaking abilities!).

These are some of the known risk factors for developing a noisy hip:

◊ If you're young (under 50), tall, and heavy.

◊ If you regularly do activities which move your hip through a wide range of motion (gymnastics, for example).

◊ Precisely where the prosthesis is positioned, and its overall alignment. Some positions are more prone to becoming noisy than others.

◊ Ceramic hip replacements are more prone to squeaking than plastic and metal.

◊ Smaller implants squeak more often than larger ones[76].

LATE HIP INFECTION

Once you have an artificial joint in place, there is *always* the risk of infection. Late hip infection refers to infections in the hip joint that occur well after your surgery, and long after your wound has healed.

There's no need to be terribly worried about this, as relatively few cases occur after one year post-surgery[77]. The rate is usually less than 2%. This can vary depending on your individual circumstances (smoking and obesity increase the risk of infection). But you must be careful and mindful of the risks for the rest of your life with your new hip, and know what signs to be on the lookout for.

If you have certain sorts of infections elsewhere in your body, the bacteria can travel via your blood to your new hip, and set up shop there. The bacteria lodge around the prosthesis, producing a "slime envelope" that protects them from antibiotics. The prosthesis and the surrounding soft tissues end up coated in these slime-protected bacteria[78].

The immune system is unable to attack these bacteria on the artificial surface of the hip replacement; the area around the hip prosthesis often has a poor blood supply, which doesn't help matters either. Apart from the risk of creating antibiotic-resistant bacteria, the infection can even destroy bone and loosen the prosthesis (septic loosening).

Unfortunately, these sorts of infections are hard to treat with oral antibiotics. The treatment generally involves having to return to hospital to have the prosthesis removed, the joint area cleaned out, and a new prosthesis put in, and weeks on IV antibiotics. *Very* unpleasant, and not something you want to do.

To guard against this rather horrific scenario, it's **vital** that you're careful about hygiene and infection. First of all, always let your dentist and doctors know that you have a joint replacement.

Infections where there is a chance of bacteria entering the blood system are the ones to be concerned out. They must be treated seriously and urgently. These include:

◊ Boils

◊ Dental decay

◊ Infected cuts

◊ Ingrown toenails

◊ Kidney infections

◊ Nail infections

◊ Prostate infections

◊ Skin infections

◊ Tooth abscesses

◊ Urinary tract infections (UTIs)

Infections like sinusitis and bronchitis aren't such a worry, as there is a low chance of bacteria entering the blood system. But basically, if you're sick and running a fever, notice burning urination, or an infection on your skin, see your doctor. They will be able to assess the level of risk, and whether you need antibiotics or not.

Over a year after my THR, I was away at the beach, and noticed a very mild burning sensation when I went to the toilet — early signs of a UTI. I was far from home, so couldn't see my GP. I went to the local hospital's emergency room. The doctor there completely agreed with my decision and the urgency of the situation, and prescribed antibiotics. **Denise, 47 Australia**

Before you have any dental, urological, gastrointestinal, or surgical procedure, make sure your specialist knows about your prosthesis; you will probably need precautionary (prophylactic) antibiotics beforehand. This includes screening procedures like colonoscopies, bronchoscopies (endoscopy of the lungs), and cytoscopies (endoscopy of the bladder).

If you're having major dental work or invasive cleaning done, it's advisable to be on antibiotics beforehand, just in case. In general, though, regular dental check ups and regular cleaning won't require antibiotics.

If you have an immune system disease (HIV, cancer, sarcoidosis, endometriosis, lupus, multiple sclerosis, and so on), diabetes, or rheumatoid arthritis, or take corticosteroids or immune-suppressing medication (eg Methotrexate, Imuran) for whatever reason, you have a higher risk of developing this sort of late infection in the months and years after your surgery, so be on the alert.

◊ Some surgeons advise wearing surf shoes when you're at the swimming pool or the beach, to reduce the risk of cuts on your feet and picking up infections.

◊ Keep your fingernails and toenails in good condition, and avoid infections, don't cut your nails too short. Do your best not to bite or pick at them. Don't tear cuticles.

◊ If you suffer from recurring ingrown toenails, your doctor may advise you to have the side of the offending nail bed(s) removed surgically. This is a minor procedure generally done under local anaesthetic in the surgeon's rooms or by a trained podiatrist, and can save you years of infections, worries, and antibiotics.

◊ Maintain good dental health. Have regular check ups, and brush and floss regularly.

◊ Wear thick gloves when gardening, especially if working with soil, compost, or prickly plants. Roses are especially known for spreading infection on their thorns.

◊ Never allow anyone to inject cortisone or other medication in or around your artificial hip, as it can cause infection.

Symptoms of Possible Hip Infection

◊ Persistent fever (higher than 37°C/100°F orally).

◊ Shaking chills.

◊ Increasing redness, swelling, tenderness around your hip/groin or around the wound.

◊ Drainage from the hip scar area.

◊ Increasing hip pain with both activity and rest.

◊ Reduced mobility.

See your doctor *immediately* if you develop any of these signs[79].

ASEPTIC LOOSENING

This is the most serious long-term problem with hip prostheses. *Aseptic* means "free from contamination by bacteria or other microorganisms". In other words, it's *not* an infection. *Loosening* refers to the prosthesis becoming loose within your bones. The general failure rate of THR from this problem is around 10% after 10 years[80,85]. Congenital hip dysplasia does also present a risk factor for prosthesis survival, so dysplasia patients have a higher chance of developing this complication[81]. Aseptic loosening is the main reason people eventually need revision THRs.

Factors that affect this are:

◊ How well your surgery was done.

◊ What sort of prosthesis you have.

◊ Whether the prosthesis been cemented in place or not.

◊ Your activity levels (more loosening with more activity).

◊ Your weight (less weight = less stress).

◊ How strong your bones are.

The longer you've had the prosthesis in place, the more likely it is to fail. Basically, all hip prostheses fail eventually, it's just a matter of from what cause. Generally they just wear out from use. If you are older, you may find

315

that your new hip lasts for the rest of your life. But, if you've had a THR at a younger age (under 50) — which many dysplasia patients do — you are likely to need a revision surgery eventually (see Chapter 4 for further discussion about this), and are more likely to develop aseptic loosening. Prostheses with plastic (polyethylene) surfaces are more likely to develop aseptic loosening [82].

There is still some debate about exactly what happens to cause this problem, but the current theory is that microscopic wear particles from the artificial joint lodge around the stem of the prosthesis, the body attacks these particles, and the bone is affected. These particles can be polyethylene, cement, metal, or ceramic, depending on what your artificial hip is made of, and how it is fixed in place. The size and number of the particles is more important than what they're made of [83]. When there are a lot of wear particles, the body can't clean them up effectively. There appears to be no significant difference between cemented and non-cemented hips, although this finding is still being debated [84].

The body's immune system sends white blood cells (**macrophages**) to get rid of these foreign bodies, but unfortunately they die when they ingest the particles, and release enzymes and other chemicals (**cytokines**) into the joint space. These enzymes and chemicals cause the bone to disintegrate. This process is called **osteolysis**.

As the bone is eaten away, the fit of the stem inside your femur, and the socket in the acetabulum, gets worse, and things can start to move around.

Signs of Aseptic Loosening

◊ Pain in the groin/hip region, with pain radiating into the thigh.

◊ Pain in the middle of your thigh.

◊ Sense of instability of the joint.

◊ Reduced range of motion.

◊ Increase in pain after increased activity.

◊ Possibly some warmth, reddening, and/or swelling over the hip joint.

An x-ray can usually show up aseptic loosening, and your surgeon will know the other signs to look for. You may be given a skin patch test to see if you have any metal hypersensitivies to the materials in the prosthesis. Chromium, nickel, and cobalt can be problematic for some people, although this is a rare complication.

Pain killers can help you cope with the pain (obviously), but there is no medical (non-surgical) treatment for this problem. Unfortunately, once again, surgery is the only solution.

Aseptic loosening takes a long time to develop, so if you have these sorts of symptoms only a matter of weeks or months after your surgery, it's more likely that you're looking at an infection. You still need to see your doctor!

DISLOCATION

If your new hip dislocates, it means that the ball slips out of the socket, and you will know about it right away!

Dislocation is more likely to happen within the first six weeks after surgery, when the joint is still recovering. There is a higher risk of dislocation if you have (you guessed it) hip dysplasia. DDH patients may have as high as an 8% risk of dislocation (1-4% in non-DDH patients). Dislocation is also more common in revision THRs (as much as 16%)[86].

Smaller femoral heads have a higher risk of dislocation. It has been shown that femoral heads 32 mm diameter and larger have less risk of dislocation than small femoral heads (22–26 mm diameter). Other risk factors include: alcohol abuse, prior hip surgery, being female, increasing age, dementia, and diseases that damage the muscles. A posterior surgical approach when your hip prosthesis was put in has a higher risk of dislocation too[87].

Mechanically, most dislocations are caused by the **impingement** of the neck of the femur against the rim of the socket. The levering action when the ball hits the edge of the socket makes the ball pop out. If the muscles around the hip joint are weak, they can't hold it all in place.

If your hip dislocates, you will know about! It is *very* painful. This is an emergency situation, call an ambulance, and get to hospital as soon as possible. Do not eat or drink anything after a dislocation, as you will probably need anaesthetics.

Sticking to your hip restrictions are the best way to avoid a dislocation. However, even if you're very careful, you may accidentally slip or fall over, which can end up in a dislocation. Don't feel bad if you end up with a dislocation, they happen to the best of us.

You will have an x-ray to confirm the diagnosis, and your surgeon may be able to pull on your leg and pop your joint back into place. This happens *after* you've had anaesthesics and muscle relaxants, in the emergency ward. In some cases you may need surgery to fix things. After your hip is repositioned, you probably need to wear a brace to position your hip in a safe position, while the muscles and soft tissues recover. It will take some weeks to recover.

If you have repeated dislocations, this may mean that your hip joint is too unstable, and you may need surgical treatment to stablise the joint.

Debbie very generously shares her experience of her hip dislocation:

Two days before my all clear to drive (six weeks post-THR) my right hip dislocated. I was off crutches and, worry-free, I went to go upstairs. On the second step there was an almighty clunk, and I was hit with the most excruciating pain (on par with child birth!). I don't know how, but I turned and sat down on the step, and then it dawned on me what had happened. My fear was that my hip had broken, and not dislocated. I couldn't move a millimetre without all hell breaking out, and pain beyond belief.

I couldn't move get to the phone, so all I could do was scream for help. Although I knew it was unlikely that anyone could hear me, I still yelled on and off for an hour until my husband John came home. He took one look at me, dropped the shopping, and came to try and move me. His face was ashen as he dialled 999. As the relief flooded through me, it was soon over shadowed by the horrific all-consuming pain I was in.

The ambulance came, and with it "gas and air". I needed it as they strapped me into a chair and carefully carried me up the garden stairs. I stayed strapped in all the way to A&E, as it was too painful to even try and move over to the bed in the ambulance.

Embarrassingly, I could not stop screaming, as our neighbours watched on. The half hour journey to A&E took forever, and every bump on the road felt like torture. On arrival I can't really remember much until I went to x-ray. I didn't care as they cut my jeans off me — my best pair! They told me my hip had dislocated and it wasn't broken. I was given morphine, but still the pain rattled through me.

A doctor came to tell me they were going to try and put it back under sedation, as there was a long wait for theatre. They tried to do it, but alas it didn't work and I have since discovered it was unlikely to have worked, as you need the deep relaxation of the muscles to reset a dislocated hip, only attained with a general anaesthetic.

John, who had watched the process, looked as if he had been to hell and back (don't ever let someone you know and love watch medics try and put a dislocated hip back in). He couldn't believe I was so calm and collected, and so obviously unaware of the brutal procedure I had been through.

More morphine and up to a ward to await theatre space. Approximately 15 hours after dislocating my hip, I had a general anaesthetic. The next thing I knew, I was awake, and the all-consuming pain had gone. Back in the ward, I was up on my feet and off to the bathroom. After a visit from the physios, I was back on crutches for six weeks. The doctor let me go home on the conditions I treated my hip as if I was on day one post-op. Very disheartening.

My hip was once again mending, and I saw my surgeon, who was very concerned about the dislocation. He was pleased to show me the x-ray with the hip back in place. He still couldn't work out why it dislocated, but warned me it could again so I had to take great care.

The fear of it happening again was far harder to cope with than the actual pain of the dislocation! Every slight click or clunk from my hip strikes a deep and real fear in me.

It was my physio who worked out why it dislocated. I had had a total knee replacement the previous year, and the muscles around my knee were still weakened. He suggested that as I went upstairs my knee moved inwards, which consequently pushed my hip to push out and dislocate. Strangely, knowing why my hip dislocated, has made it easier for me to cope with.

18 months after my dislocation I still get the odd nightmare about it, and I still break out in a sweat if I bend too far or turn awkwardly. If John catches me weeding in the garden or just standing in the wrong position, he will tell me off. I know he is right, and totally against my usual character, I will do as I'm told! Both of us have been traumatised by my hip dislocating, and have had to come to terms with the whole horrific experience.

When your hip has dislocated in time it physically heals, however the fear and terror of it happening again doesn't. I saw my surgeon recently and now my left hip needs replacing. Although obviously one doesn't look forward to any operation, the thought of the actual operation doesn't worry me too much. I know the nagging draining pain from an arthritic hip disappears immediately after surgery. However, the thought of another dislocation, however unlikely, haunts me and fills me with real trepidation. Logically I know it will be fine, but the fear of dislocation changes me from a fairly together and logical person to a fearful and truly scared person!

Having spoken to other people who have had dislocations, I have discovered that they have experienced the same clunks and clicks I experience when I'm doing something silly, like bending wrongly whilst clearing out the cat's litter tray. One old guy told me that every time he puts his right sock on he doesn't know if it will click back in or clunk out! I have since discovered from my surgeon that the clunks are made when the hip is going back into the socket.He also said a fraction of a millimetre further could just as easily clunk out. Scary hey!

I now let my husband do the cat's tray, and I leave the weeds in the garden to stay put and flourish. Dislocated hips need to be avoided at all times, and even if I get irritated if my husband tells me to stop doing something because I am bending in the wrong way, I just need to think of my hip dislocation, and then I am only to pleased to be warned. In this case, prevention is 100% better than cure!

In conclusion, all I can say is don't do anything which may put your new hip in danger of dislocating! Although it is good to walk and exercise after a THR, it is much better to be slightly cautious, and take a little longer to do things than to risk the chance of a dislocation. **Debbie, 52 UK**

Please don't be too worried about the information on complications in this chapter. Most people have no problems after a THR or PAO, and these procedures give them a new lease on life. Most of the time, you will have no problems adjusting to restrictions, and honestly, most of the serious complications we've covered here are rare.

Chapter 16
Back to Normal Life

There is a point after your surgery, it may be a matter of weeks or a few months, where things start returning to normal. Your carers go back to their own lives, your partner goes back to full time work, your staples or stitches are out, and you can mostly get by without a walking stick. People stop thinking about your hip (even though you don't).

But when is it okay to start doing 'normal things' again? We are going to cover what we feel are important aspects of normal life that warrant discussion, including driving, sex, alcohol, returning to work or studies, shopping, public transport, exercise, and a few other bits and pieces.

Alcohol

There are no hard and fast rules as to when it is okay to start drinking after major surgery. If you fancy a glass of wine to celebrate as soon as you get home from hospital, well, you can. The main thing is alcohol interacts with lots of medications, and it's not a good idea to drink when on strong pain killers; in fact, in some cases it can be dangerous.

NSAIDs can affect the lining of the stomach and lead to nausea. In the worst cases, they can cause stomach ulcers. This effect may be worsened or accelerated if alcohol is taken alongside NSAIDs.

There may be times when a glass of your favourite tipple may help you relax a little bit, and it can sometimes help you sleep. Just remember to drink safely, and in moderation.

BEING SENSIBLE

If you are going out with friends and planning to drink while using crutches or a stick, or even when you are not, it's important to recognise your limitations. Alcohol affects coordination, and getting drunk makes most people unsteady on their feet. If you are not 100% confident in your stability, or still on crutches or a walking stick, it would be wise to bear this in mind when deciding how much to drink. The last thing you want is to fall onto your hip, and possibly dislocate your new joint or damage your PAO!

Caring for family

Hopefully your family have been looking after *you* for most of your recovery; however, there will come a point where you will feel able to do more for yourself, and for them, and get back to normal family functioning.

Don't do too much too soon, this will only make you exhausted, or set you back. The key is to do little chores and tasks, with lots of rest in between. It is amazing how much an achievement making dinner, or putting on a load of washing is, the first time you do it after surgery!

Running around after little children is also a great strain on your new hip, and you may still be very exhausted for quite some time. Now is not the time to host big birthday parties, or volunteer at your child's school. That can come, but later.

Driving

If you live a city with good public transport systems, or where taxis are readily available, it is possible to get by without a car. However, for the majority of people, a vehicle of some sort, and being able to drive, are an important part of your independence.

There will be a period of time after your surgery when you will not be able (or allowed) to drive. To be honest, you probably won't feel like driving anyway, so this may not be too much of an issue.

Another thing to consider is what type of car you have. If you have an automatic car, and you surgery was on your hip that doesn't work the brake or accelerator, then once you are off medication and feel able to do so, you should be able to drive relatively soon. If you drive a manual car, which requires both legs to be working well, you will have to wait longer.

Type of surgeries and post-operative protocols vary from surgeon to surgeon, so follow whatever advice you have been given on driving. We cannot emphasise the following point enough — if your surgeon has said you must not drive for 'x' weeks (regardless of what has been done or to what side), they will have documented this advice.

If you drive within before that period of time has elapsed, and have an accident, you **will not** be covered by your insurance. This can have significant financial and legal implications for both you and your family, and also the other driver and their family, so please just be sensible, however frustrating it is. Usually there are reasons behind these rules.

If you really feel you are ready to drive sooner than your surgeon has said, you can always ask them to document that you are allowed drive sooner (if they agree).

Getting into and out of a car can be a bit tricky post-op, and in the short term, but long term it shouldn't cause you any issues, unless you have permanent hip restrictions. We cover this in Chapter 15.

Driving for many people is a big step, part of regaining your independence. It can feel so exciting the first time you go out by yourself in the car — doing something so *normal* again. It may sound obvious, but is a good idea to start with small trips, and build up to longer journeys. If you absolutely have to do a long journey sooner, take frequent breaks so you can stretch and rest. You may find your hip can get stiff easily while driving, at first.

PARKING

You may find that trying to twist and squeeze out a vehicle in a narrow space is difficult for some time. If this is the case, follow our advice on disabled parking in Chapter 6. It may require a bit more planning and effort, and in some cases more walking, so bear this in mind when you make that first solo trip out. It is easier to get out of the car if you swing both legs around to the side of the car seat, and stand up once both feet are on the ground, rather than trying to swivel out sideways on one leg. (See Figure 15.4 on pg 301).

Exercise

For many hippies, getting back to some form of regular exercise is a major goal. However, it is important not to set yourself *unrealistic* goals. You have just had major hip surgery, after all! Be gentle on yourself, and forgiving. It takes time.

Be guided by your physiotherapist, they can help you to rehabilitate your hips back to sporting standards (if that's what you're after). However, this it isn't an option for many people relying on national health services. In this case, some of the following information may be helpful to you.

Pain is a remarkably accurate tool for limiting your activities. If you are doing something, and your hip starts to hurt, it is a sign to stop.

Finding exercise that you can do pain-free is the ideal situation. As your muscles get stronger and used to exercise again, you can progress on to other activities. If your surgeon has advised you not to do a certain type of exercise, however hard it is to hear, it is usually for a good reason, so please follow their advice. The most common restriction after DDH surgery is no running or other high-impact activities such as jumping, basketball, soccer, hockey, and karate.

We discuss good hip-friendly exercises in more detail in Chapter 3.

Pregnancy and childbirth

It is definitely possible to be pregant and give birth with DDH, pre-surgery, and post-surgery. There are many women out there who have successfully conceived, carried, and delivered children despite having hip dysplasia.

PREGNANCY

The fact that you are reading this book means that you are aware of your hip condition. If you have decided to wait for surgery on your hips, or have been told that you are not appropriate for hip joint preservation surgery, but are too young for a replacement/resurfacing, you may not want to put all other aspects of your life on hold.

If you have had surgery on your hips, you may have no problems with them during pregnancy whatsoever, or you may have ongoing problems.

Your body undergoes many changes during pregnancy aside from the obvious weight gain; these changes may have an impact on your hips (even if you do not have any problems at the time when you conceive). Some hippies find that their hip pain decreases in pregnancy, while for others it gets worse. In other words, it is really hard to predict how things will go for you until you try it. But the main thing is, even if your hips hurt a lot during pregnancy — it's not going to affect your baby's health, so it's quite safe in that regard, you may just have to be stoic.

Planning your family and discussing your hip issues with your midwife, and the doctor in charge of your ante-natal care, is important. They need to be aware of the situation, and can then be in a better position to advise you. It's a good idea to have a hip x-ray *before* falling pregnant, so your obstetrician can measure the dimensions of your pelvic outlet and assess your hip anatomy, to see if a vaginal birth will be possible for you or not.

GIVING BIRTH

Whether you have had DDH surgery or not, there are a few things to think about. It is a good idea to have a trial at the place where you are going to give birth, to work out what positions you are able to get into and are going to work for you (from a hip perspective), before you are in labour. It probably isn't wise to opt for a home birth if you have dysplasia, as your abnormal anatomy can give rise to unexpected problems during labour.

Because my first pregnancy was a very welcome surprise, there wasn't an opportunity to do a hip x-ray to see the state of play. I was (ridiculously) adamant about having a home birth, and the local home birth doctor agreed with me — in retrospect, he *should* have refused, knowing my hip history. My pelvic outlet is actually quite deformed as a result of my dysplasia and childhood hip surgeries, and no baby can ever get out that way. I ended up with a 32–hour obstructed labour, with 24 hours of labour at home, and eight hours of pretty dreadful labour in hospital while I waited for an emergency Caesarian. Thankfully, my son wasn't distressed by the process. Two years later, my daughter was born by planned Caesarean. **Denise, 47 Australia**

In one of the few studies that have been done evaluating sexual activity, pregnancy, and childbirth after periacetabular osteotomy, it was shown that the majority of women will still be able to deliver a child vaginally[88,89]. The reasons for Caesarean section were the same reasons that women with no hip dysplasia undergo Caesareans. However, every person and case is different, and needs to be considered individually. Be guided by the medical experts looking after you.

TIMING

When to have your baby? Pre-surgery? Post-surgery? What if you're already pregnant, and then diagnosed with hip dysplasia? Lots of women undergo major hip surgery with children at home (of all ages), you just need to take this into account when planning your hospital stays and recovery at home. You need to consider when you're thinking of having surgery (PAO soon, or THR later for example), what the surgery is, and when you're wanting to have children. Reading Chapter 4 can help you to be informed about the recovery times involved with each surgery.

Obviously, it is easier to have surgery, and get over it, if you *don't* have to look after young children at the same time, but on the other hand, children are so inspiring and full of love — and often even helpful — that they may motivate you during your recovery. There is more about this in Chapter 9.

My experience of being pregnant after my THR is a very positive one on the whole, I would say. I got pregnant nine months after my THR (my OS had advised waiting at least six months). I experienced no problems or pain at all in the replaced hip throughout my pregnancy. In fact, my non-replaced hip also benefited with all the relaxin in my body and didn't hurt at all either, where before getting pregnant it was niggling and sore most days.. There are two factors that helped though — one is that I didn't put on loads of weight — only 16 lbs up to week 38, then another 6 lbs in the last four weeks, and secondly, I walked a lot — at least two or three times every day, and I think that really helped.

My major concern was delivery. I had been reassured by my ortho-consultant that I should have no problems with trying for a natural delivery, and also from an obstetric consultant, but I had my doubts. The reality was that I was two weeks overdue, was induced for three days, and ultimately ended up with a c-section because my labour failed to progress. Baby was back to back and his head didn't engage. Whether that had anything to do with my wonky pelvis I'll never know — but my gut feeling was that it did. Labour pains were all in my replaced hip side and lower back because of the baby's position. I was worried I'd be left with residual pain, but I don't have this at all, which is a big relief.

The THR hip is also holding up well with all the activity that goes along with a baby — lifting, carrying, bending and so on! **Deirdre, 36 UK**

Public transport and events

Public transport can be an "adventure" post-op, especially when you're using mobility aids (crutches, a wheelchair, or a walking stick). It's a good idea to plan your journey in advance, taking into account any stairs, escalators, and the availability of lifts. Using a backpack/rucksack means your hands are free to use crutches, or hold onto rails safely.

Remember that you have to be able to do *everything* yourself, especially if you are travelling alone. Do not expect that members of the public will help you, or even offer you a seat. In major cities, there may be special "disability-friendly" maps available online that detail how many stairs there are at stations, the presence of lifts, gap between the trains, locations of disabled toilets, and so on. If using a taxi service, inform them when you are making the booking about any mobility issues; most large companies will be able to provide a suitable vehicle, if they are given enough notice.

It is sometimes easier when you go out if people can see there is something obviously wrong. If you are using a stick or crutches, people can see there is a problem, and are more likely to get out your way, or be more understanding when you ask for a seat.

There will be a time when you don't really need a mobility aid, but still get tired with long journeys, or walking long distances. It's worth keeping a walking stick or a single crutch to hand for times when you've got a lot of walking to do, and are likely to get fatigued, even if you don't desperately need it for support, and aren't limping any more. They're a bit of a pain to keep using, as you know, but there's nothing better as a visual aid to onlookers, and can stifle all sorts of unwanted comments, as well as helping when you're worn out. It is possible to buy fold-up walking sticks.

I used my crutches while going out in London on the Tube for quite a bit longer than when I had stopped using them at home. It meant people gave me a wider berth, and usually meant I got a seat. If not, when I asked for a seat, was obvious why. **Sophie, 29 UK**

AIRPORTS

If you are flying anywhere and anticipate needing extra help, it is definitely worth organising assistance, rather than trying to manage on your own. You can inform the airline when you book the flight, or at any point up until you fly — even at the check-in desk is not too late.

Airlines can provide you with a wheelchair, allow you to jump queues, speed your way through security, and board the place early ahead of the rush, so you can take your time getting on, and getting settled in your seat, with help. Well, there have got to be *some* perks!

This also means that there will be assistance awaiting you when you get off the plane at your destination, and if you are travelling alone, they usually help you with baggage collection too. We discuss air travel in Chapter 14.

Even if you're recovered from a PAO or THR, you can still experience fatigue or pain when navigating through large places (airports, malls, and so on), or when carrying luggage. I strongly recommend arranging to have a wheelchair provided.

My nephew (who has had a THR) was traveling and planning to carry his guitar as well as his carry-on luggage. I suggested he request a wheelchair from the curb to the gate so he wouldn't end up incapacitated, as he still experiences pain, and walked with a slight limp at the time.

The wheelchair worked out so well — it left him with full energy and pain-free, so he could actually tolerate the flight and then enjoy his trip. He does it every trip now!

I should add that at first he was resistant because he's a macho kind of guy, but I was insistent and eventually convinced him to give it a try. I used to be a travel agent, and explained how easy it is — just request it by using the toll-free phone number for the airline even if you booked your tickets online.
Brenda, 48 USA

There is a possibility that if you have metal in your body (PAO pins, metal hip replacement) it *may* set of the security machines at the airport. This may require you to be searched with the hand wand, or by a same-sex security officer. They are looking for weapons, and it will quickly become obvious you haven't got any (we hope!).

Simply explain that you've had hip surgery, and have an artificial hip or metal screws. The security screeners can be unpredictable too — in some places you may set off all the detectors, and in other airports you may walk through without a ping. If this is something you're particularly concerned about, simply carry a letter from your doctor.

Bri has flown several times since surgery — no security issues at all. We were told by her surgeon that because the pins are stainless steel they won't set off the detectors. **Cynthia, 47 USA**

I went to Köln, Germany, and nothing happened about my PAO pins on passing through the detector. **Alvaro, 38 Spain**

Sometimes my hip resurfacing sets off the alarms, sometimes it doesn't. It's never been a problem — I just make sure I get rid of anything else (keys, coins and so on) to speed things up. As soon as the alarm goes, I just tell them I have a hip replacement (easier for them to understand than "resurfacing"). Then they either pat me down, or make me stand in one of those silly "x-ray vision" thingies. I have never bothered with a doctor's note, as anyone could forge one, so they are meaningless. I just figure if it comes to it, I'll show them my backside (the scar, that is …). **Freja, 45 UK**

My 25-year-old nephew (who has had a THR) carries a card from his surgeon that I think comes from the hip prosthesis manufacturer to show at airport or courthouse security, or any other location that uses metal detectors for security. **Brenda, 48 USA**

I travel quite a lot for work and my THR (ceramic on ceramic) has set off the metal detectors at airports all over Europe. It has never been a problem, as most airports then employ additional screening with a handheld wand combined with a pat down over the hip. They are quite happy when I tell them I have an artificial hip — it's pretty obvious that I do when they wave the wand over the left hip. It doesn't happen 100% of the time, however (although always in the UK). A colleague has suggested that sometimes they don't seem to go off if I walk quickly through the archway, but I'm not sure. I also keep a picture of my x-ray showing the THR on my phone in case I ever have problems or can't communicate in English, but never had to show it.

The only slight difference was when travelling recently through T5 at Heathrow in London, where I had to go for secondary screening in a full body scanner. It was kind of embarrassing, as I had to hang around and wait for someone to take me along to the scanner, but not a big deal really. **Deirdre, 36 UK**

CONCERTS AND SPORTING EVENTS

If there's a sporting event or concert on that you really want to attend, but are still on a walking stick or crutches, give the venue a call. Many of the big venues have special seating areas for people with disabilities, and even special ticket prices. Your carer may even get a free ticket! You may need some documentation, such as a letter from your doctor, but it's definitely worth checking out.

Sex

This is a subject that can be very difficult to ask your surgeon about. It's an extremely important topic, yet one that is avoided at most consultations because of embarrassment. It's hard enough to discuss with your partner! Most of the medical team looking after you will have no idea what it's like to try to have sex with hips that cause so much pain and difficulty. The problems you may encounter include pain, stiffness, catching in the joint, and loss of libido. It's hard to get excited about doing anything intimate when it causes pain, or you're worried it's going to cause pain, or injure your hip.

In one study, female PAO patients were surveyed, and asked to rate their satisfaction with sexual activity after their surgery. 40% reported a change in the frequency of sex post-op, 46% had changed positions post-operatively, and 25% of patients reported changes in satisfaction of sexual intercourse. None of the patients who reported more frequent sexual intercourse post-op had retroversion (hip socket tilted backwards)[88]. Another study showed that young women had more satisfaction with sexual activity after PAO, because of the reduction in hip pain post-op[89].

SEX WITH LONG-TERM PARTNERS

Having an understanding partner is truly wonderful, and you need to remember that this situation is difficult for them too. Discuss how you feel with them, and your expectations. Sex definitely *is* possible, even when your hips are sore (either before or after surgery). Positioning is all-important, and it is likely to be a case of trial and error, working out what works for you while getting sexual satisfaction.

Be aware that after major surgery in the hip and groin region (especially post-PAO), you can get very swollen "down there" (labia and perineum). An ice pack may help matters.

Also remember there are plenty of ways to be sexually intimate with someone, without necessarily having full intercourse. This is particularly good to remember in the early post-operative period. Be inventive!

After a THR, sex is possible as pain allows, as long as hip restrictions are adhered to (see Chapter 15). After other hip surgeries, sex is usually allowed as pain allows, however, in those where osteotomies have been done, it might be wise to wait until the bones have started to mend a bit (usually after four to six weeks), though there is no evidence for this (but then we can't imagine quite how a study into this topic would be carried out!).

Quite honestly, most people find that there is no real desire for sex until after about six weeks, when the pain has settled. Each individual and couple is different, so finding out what works for you both is important.

SEX TIPS AND TRICKS

◊ Use pillows to support your pelvis or your legs ,so that you do not have to use your muscles all the time.

◊ When on top, doing it very close to the edge of the bed means that your leg can hang of the side, this can be useful if flexion (bending your hip) is painful.

◊ Spooning is good, lying on your side behind or in front of your partner, as your legs can be close together. Keep a pillow between your legs if you're post-THR, and only lie on your side if allowed by your surgeon.

◊ It sounds strange, but doing an exercise "warm up" with some stretches before sex can help loosen your muscles.

◊ Be sure you're *really* up to it before resuming your sex life, otherwise your partner may think things are back to normal too soon!

Our dear contributors have been very generous in sharing their experiences candidly. Some names have been changed for privacy:

The pressure of sex on my hips was at times uncomfortable, I'd have to be on all fours; doggie style was most comfortable for us both. **Lisa, 40 USA**

Besides being scared to have sex right after the surgery, for fear of pain and/or injury, it took me a long time (at least a year and a half) to be able to not feel significant pain or stiffness during sex. For over a year, the "girl on top" position would result in me having to change positions after a while (~15 minutes) or getting stuck there (unable to "dismount!"). Missionary was easier but it was difficult for at least the first six months to hold my legs up in the air! **Jill, 20 USA**

After my wife had her PAO I was very worried about sex at first, as I was so scared of hurting her. However, with practice, we gradually found ways that were comfortable for both of us, though we still have to be careful. **Adrian, 32 UK**

Just like everything else with my hip, it crept up on me, really. I first noticed it if I was on top — after a few minutes my hip got really stiff, so I stopped doing that! Then opening my legs with the knees bent (missionary position) became gradually more and more painful too. Sex became less and less fun, as I just got more and more preoccupied with the pain (and trying to avoid pain). I just felt so tired all the time too.

All the stuff that came with my bad hip (using crutches, getting clumsy and slow, only ever wearing comfy, slouchy clothes, being on guard against pain constantly) just made me feel years older. I had a real problem with just not feeling that attractive.

I was having a rubbish time in bed (not my lovely other half's fault — he tried his best!). I must admit I felt quite weepy and sad about it at times. In the last three months or so before my surgery, we just stopped doing most things, as it made us both too miserable. **Freja, 45 UK**

This hip operation has made things very uncomfortable for me. After my left PAO, it took about five months before I was starting to feel more comfortable, and I was able to get into a position that gave me pleasure, and wasn't too uncomfortable. To start with my hip just wasn't flexible enough, but it did start to improve.

Then I went in for my right PAO, and also suffered a stress fracture, which between the two caused huge discomfort when trying to do anything sexual. I was still struggling to be comfortable when I fractured my ischium (eight months later). Well you can guess — that really didn't help! It is only more recently in the last month or so, that I can honestly say that I whilst in the process of "doing it", I am actually thinking about pleasure, rather than pain in my hips. Prior to this, each thrust was like an eternity and I was just wanting my husband to get on with it, and get it over with as quickly as possible (in the nicest possible way — I did want him to enjoy himself!). I will say that it wasn't a totally displeasurable experience, but it was also still quite uncomfortable! The saying "pleasure and pain" comes to mind!

333

I also don't think my husband really understood the discomfort involved, particularly with a double PAO (and I am not blaming him here, it's just a fact). I have found that I have had very little desire to resume activities in this department because of the pain I have and the pain I anticipate having. Having said that each time, it has been getting a little easier and little less painful, but the desire still isn't back to what it should be. My husband has more recently started to understand the pain I have been in, when he realised how easy it was for me to fracture this last time; and Mr J confirmed that if my more recent fracture didn't settle, that I would be up for a bone graft and plate.

The whole thing has made me more hesitant about wanting sex — a lot of it is psychological, but it has made me feel vulnerable and unsure. Difficult to describe really, but I'm a little less outgoing and more reluctant. **Annick, 47 UK**

Before PAO, everything in the bedroom was fine, just clunky when I moved in certain positions, no pain though. After surgery my range of motion is still limited — one year post-op, sex is more difficult and not as relaxing as it should be. Positions are more limited now due to pain in different positions.
Lea-Anne, 40 Australia

Emotionally I don't feel good, due to the pain and I don't feel attractive because I have put on weight. Physically, obviously my hip condition renders some positions extremely painful or impossible. This limits positions and variety. I find if we change positions that are okay quite regularly, my hips don't feel quite so stiff and painful. I do find every position painful, but some are worse than others. This also means that my partner tries to be very careful not to cause me pain, but that really frustrates me, even though he is just taking care of me! Taking medication also reduces sensitivity in areas, which is ultimately not very satisfying. **Olivia, 25 UK**

It didn't really affect me, although I did find my sex drive had decreased quite a lot over the surgical and recovery period, but I put this down to anxiety of not wanting to break my new hip and the effects of medication! **Rhianna, 28 UK**

334

There is also a range of times that people wait before resuming their sex life.

My surgeon allows sex to resume after six weeks. We went pretty gingerly about it at first and even now at nearly three months out, it is still uncomfortable; the best position we have found is for me to lie on my non-op side and grasp my operated leg and scoot to the edge of the bed so he stands and enters me from the edge of the bed, this way I can cradle my leg and control the angle, and he can keep his weight off me and control the thrusting. **Lisa, 40 USA**

I was still on crutches when I resumed sexual activity, which was exciting in and of itself! It was about five and a half weeks post-PAO. **Jill, 20 USA**

We had a bit of a cuddle within about two weeks, but waited almost three months to try out my new hip. I did not expect anything too amazing, it was a bit scary at first, but even that first time, we could both tell there had been an improvement, so it made us hopeful! Within about a month, things had improved considerably — what a difference a new hip makes! **Freja, 45 UK**

After my left PAO, we made love four weeks after, but after my right PAO and fracture it was eight weeks (Christmas Day!), but it has been very erratic this year to say the least. **Annick, 47 UK**

I felt like a stallion just the night after my PAO surgery, but my wife calmed me down. We have resumed our sexual life in the fashion we can. **Alvaro, 38 Spain**

We waited til three weeks post-op. **Lara, 40 USA**

We waited three months just for precautions and confidence. I wasn't going to damage my new hip!! **Rhianna, 28 UK**

We resumed our sex life at around four or five months post-PAO. But I was total non-weight-bearing for almost three months. **Lea-Anne, 40 Australia**

Some more advice from our intrepid hippies :

I have to admit it can be pretty unromantic, those first halting stumbling steps while you position, reposition, reposition, and finally end up having your husband pretty far away from you, only connected by one organ. Thank goodness he's my best friend, and we've been married for 17 years, so we can just roll with the changes, and laugh at the awkwardness later. **Lisa, 40 USA**

If you resume while you're still on crutches, be sure you trust your partner, as he'll probably have to retrieve your crutches for you afterwards. Wouldn't want someone to get mad at you, and leave you stranded! I found that sex on a couch or futon in missionary, with the op leg raised up/foot resting on the top, was the most comfortable. **Jill, 20 USA**

My only advice is to keep trying different things, don't stay in the same position for too long, and try and have a laugh about it together. **Olivia, 25 UK**

Like any aspect of a relationship, communication is the key. Strangely, I felt that even the worst times with my hip brought us closer together. Patience and a bit of compromise — on both sides — is also very important. **Freja, 45 UK**

I certainly don't feel that the doctors/hospitals help on the sex agenda. Absolutely nothing was said to me regarding when I could resume a sexual relationship — nothing! So ask your surgeon about it, and talk to your partner about it before your operation, and after. I think it is just so important that your partner understands that you probably won't want to do anything for quite some time, and that this is actually okay — it doesn't mean you don't love him! **Annick, 47 UK**

Don't try too soon, because once you are successful even if it's uncomfortable or a bit painful for your hip/leg, your partner will expect it to continue! **Lea-Anne, 40 Australia**

We found that spooning, with him behind me was super easy, as I didn't have to abduct or rotate my hips at all. **Lara, 40 USA**

SEX AND SINGLES

Being open about your condition is really important, especially with a new partner for the first time. It is far better to explain the situation early on, than to find yourself in pain later, or stuck in a position you cannot easily manage. If you have scars from childhood or adult surgeries, the topic is bound to come up sooner rather than later!

If you find yourself in position that suddenly becomes difficult or painful, it can be awkward and sometimes embarrassing to say something, and then get yourself out of the position. It may be even more so if you then have to explain how you got this sudden worsening of your pain to a medical professional!

Hopefully, whoever you are with will be understanding and supportive. Remember, they may be incredibly upset that they've done something that has hurt you, especially if they're a new partner who doesn't know your medical history. With regards to the medical profession, believe us, they will be professional about it, remember — doctors have sex as well!

A topic I was never given any advice on when I had my hip placed many years ago, as a young woman, was sexual relationships. I feel this is really sad, as hip problems have a profound effect on this and it's not acknowledged, leaving it up to the individual to cope as best they can. It certainly impacted on my ability to form relationships as I felt it was such a sensitive thing to discuss with a new person and often easier not to try for relationships as I felt inferior as a woman (in that aspect only).

I am now divorced and have a lovely partner of nine years, but I know that he is often scared to come near me in case he hurts me, especially after my recent knee replacements. **Teresa, 54 UK**

Shopping

Figure 16.1:
Secure web page
lock symbol
(circled).

Getting out to the shops can be really difficult when you have impaired mobility. It's not possible to carry anything if you're on crutches, for starters! Planning is key to any trip out when you're still recovering. Try to anticipate any difficulties, this can make them easier to deal with when they crop up. You may well need someone to come with you to the shops initially, to be your "hands".

Modern shops or shopping malls may have disabled changing rooms and facilities, which can make things easier. If you are shopping in older buildings, they may not be as well equipped. Look for things like elevators to help you get around. Escalators can still be a bit dangerous if you're on crutches. Be aware that many disabled toilets are *low*, which can be a problem if you have hip restrictions post-THR. You will probably need to master the "hovering" posture, use incontinence pads, or use a urination device (see Chapter 11). Men, you have a natural advantage here!

The internet is a great boon to anyone who can't get out to the shops easily — have things delivered to your door! The web has allowed anyone with a computer and an internet connection to shop online. Obviously this is great for groceries, books, music, DVDs, craft supplies, clothes, seeds, plants, and a ton of other things. If nothing else, it's a fantastic way to do all your Christmas shopping without stepping into a crowded mall!

For items like clothes and shoes where you might want to try them on before purchasing, it's a bit trickier, but not impossible. Clothes and shoe shops that sell online generally have extensive sizing information on their sites, and customer friendly return or exchange policies.

Many people are anxious about shopping online, but from our experience it works well, and is quite safe. Most online stores accept phone orders as well, if you're worried about online security.

The main thing to be sure of is that when you're entering any personal, banking, or credit card details onto payment pages, is there should be a tiny padlock symbol somewhere in your browser window, generally in the upper or lower right corner of the browser's **menu bar** (i.e. *not* within the web page itself— any shop can add a picture of a padlock to their web page!).

338

See Figure 16.1. This padlock symbol shows that you have an encrypted (SSL) secure web connection, and you can be sure that any information you send will be encoded and safe.

If the padlock icon isn't there in the browser menu bar, ***don't use the site***. And while we're on the subject — never *ever* send your credit card details to anyone via email.

Pre-THR I did my supermarket shop online so I didn't have to walk round the supermarket, and I avoided going to shopping malls. Now, of course, I take great delight in parking in the furthest away space from the door, and striding across the car park on my new bionic hip! **Deirdre, 36 UK**

With the wheelchair, I was able to spend hours shopping for groceries, clothes, household items, and so on. And it was so nice to get out of the house! If you can't rent a wheelchair, I would recommend using the store's wheelchair or scooter. A lot of stores keep them near the entrance. When I used a store's electric scooter, I would put my crutches in the basket in the front of the scooter, so that people would know that I was actually disabled and not just playing around on the scooter. I rarely went on shopping trips on my crutches during my non-weight-bearing weeks. It is difficult and exhausting. I usually had to sit every few minutes when I did go out on just crutches. I'm so glad I rented the wheelchair. **Melanie, 29 USA**

My THR was six weeks before Christmas. I did all my Christmas shopping online, and had gifts delivered to the house. It was so nice, that I've done it ever since! **Denise, 47 Australia**

Shopping Tips and Tricks
Planning, planning, planning!

◊ In a supermarket, use a trolley instead of a basket. This means you are not leaning to one side, or carrying anything heavy.

◊ Ask for help from shop assistants, many are more than happy to lend a hand, or help you carry a basket round a supermarket if you are in a wheelchair, or on crutches and alone.

◊ Rucksacks/backpacks and bum/hip bags mean your hands are left free.

◊ Give yourself a certain time that you will be out and stick to it, so that you don't get too tired.

◊ Don't try to do too much, or buy many things in one trip, keep it short and simple.

Some shops have special carts for handicapped people:

I have one tip — take the electric cart for handicapped people, and roll in it. Even after you are off crutches, if the pain is still there, take the crutch with you to the store so people don't ask why you are using the handicapped cart, but take the cart. It helps tons! Just make sure it is charged. Sometimes people would ignore me (or maybe I am just lucky), and it is hard to reach things on the top shelves, so the "grabber" tool may be helpful as well.

If the cart is not available (although grocery stores usually have it, and I specifically didn't go to the ones that didn't), and the choice is between wheelchair vs. crutches, I would rather go on crutches. Wheelchair made my hands hurt and they got very tired, and I almost got blisters from rolling it. Rolling it in leather gloves was the most convenient, but my personal experience was to avoid the wheelchair unless there is someone to roll me in it.
Arpine, 28 USA

Work

The type of job you do, where you work, whether you can work from home, and your entitlement to sick pay are all factors in determining when you can return to work. The type of operation you have undergone also has an impact; ask your surgeon how long they recommend you take off work (or studies).

If you can work from home and do a job that involves a lot of sitting, you will probably be able to return to work far sooner than someone whose job requires them to be very active. You also need to take into account how you will get to and from your job, as well as being able to do it well. A phased return to work is something that is definitely worth considering, no matter what job you do (including self-employment or working from home). You will get tired very easily at first and, while you may feel fine at home resting, once you start to work, it is amazing how quickly you can fatigue. Start with a few hours a day, and gradually build up. This will result in you being more productive while you are working, and ultimately a more successful return to work.

If your work has an occupational health department, you can arrange to have a meeting with them prior to returning to work. They may be able to support you in returning to work, or make recommendations to your employer. Employers may be able to provide you with special chairs to make sitting at your desk more comfortable, and other adaptations. If you have had a THR, remember to stick to your hip restrictions while at work.

The is more detailed information in Chapter 9 about working with your employer.

Self-Employed

If you are self-employed, it's really important to make sure you are well enough to return to work. Going back to work too soon may set you back, resulting in a greater loss of earnings, than if you had taken more time in recovery. It's important to let your clients and customers know that you are either off work for a time, or only working part time.

When you're your own boss, you do have the power to close the business for a month or two. It can take a while to get work done in advance, financial records up to date, and orders and commitments fulfilled before your surgery, but it will be worth it. If you have income protection insurance, it should provide your with your average income, during your surgery and recovery periods.

Remember to update your phone answering message and web site with a notice saying that the business is temporarily closed. Post a reopening date (which you can always push back if you're still struggling by then). Hopefully your long-standing clients will be only too understanding, and happy to delay projects until you're available.

If there is work to be done, or orders to be shipped out, that absolutely have to be done during your recovery, see what help you can enlist from friends or family, or even look at bringing in some professional help, either a friend who works in the same field, or from an employment agency.

You're hopefully at a point now where your dysplastic hip or hips are slowly becoming less of a focus in your life. You've got through all that surgery, you're well and truly on the road to recovery, and there are days — weeks even — when you don't think about hips! Well done! You've come so far, and things are hopefully going to keep getting better from here on in.

Chapter 17
Finding Help Online

It can be surprisingly difficult to find reliable facts and details about hip dysplasia in adults online. For starters, most searches seem to come up with information about dysplasia in dogs and babies! In this chapter we share with you the online information sources that are available on DDH, including what we found useful on our own hip journeys. The internet has changed the way we access information. It is also increasingly a way of accessing support. We also show you how to access and assess good medical information, and other hip dysplasia support systems.

Online support groups

There are several online groups that focus on hip dysplasia in adults. The main ones are **HipWomen** < health.groups.yahoo.com/group/hipwomen >, **Hip Chicks** < Hip Chicksunite.ning.com >, and the **Adult Hip Dysplasia Group** on Facebook < www.facebook.com/groups/50910135532 >.

One of the biggest appeals of these groups is the social contact with other people with DDH. You can, of course, talk to your friends and family, but however sympathetic they are, they can't *really* understand what you are going through unless they have hip dysplasia too. You *might* know someone in person who also has hip dysplasia, but it's not that likely. It is an amazing relief the first time you chat to people about what you're going through, and knowing they really *get it*, because they've been there too.

Online groups can be a good a source of information too. As we have previously discussed, surgeons who specialise in hip dysplasia are not that common, and finding a good one can be even harder. This can particularly be an issue if you live in the USA, as finding a good hip surgeon is usually your responsibility. As a member of an online group like HipWomen and Hip Chicks, you can ask the other members about their surgeons, and find recommended surgeons in your area. It's also really helpful finding out what other patients have thought of their surgeons.

Other ways in which the online groups can help you include:

◊ Reassurance about pain and other symptoms.

◊ Help getting through the stress of diagnosis.

◊ Dealing with complications.

◊ Discussing what surgery feels like.

◊ Ideas for pain relief.

◊ Listening when you need to vent.

◊ Coping at home.

◊ Delving into health insurance problems.

The list is even longer, as you can discuss anything you want! It is also a great place to broach those slightly sensitive subjects, like sex, that are just too darn embarrassing to ask your surgeon about!

Do keep in mind, though, that while members are very experienced and helpful, generally they are *not* doctors (Dr Sophie is the exception to the rule here!). So do be careful about what other hippies advise, especially with things like physiotherapy exercises, and medication — these are areas where you need an expert's advice for your particular situation.

HIPWOMEN

www.health.groups.yahoo.com/group/hipwomen/

We thought we would put this group first, as this is how we met each other. We each came across the group in different ways:

Sophie: I was working at UCLH (University College London Hospital), where a lot of surgery on patients with dysplasia is done. As part of my job, I had a list of all the orthopaedic patients in the hospital, including what surgeries they had undergone. I had already had my left PAO, and was due the right one in April 2010. I used to go round and chat (outside my professional role) to the PAO patients. I got a lot out of this, and found they did too. One told me about the Yahoo group HipWomen, so I joined, and have subsequently gained so much support from this fantastic group of people.

Denise – I was on another online community (Ravelry.com) which included a very small group for women with DDH; there were only about six of us! One of the members told me about HipWomen. In the years since I joined, I've really appreciated being a part of a community of women who have been through similar things to me, and have been hugely supportive. I only have one friend in 'real life' who also has dysplasia, so HipWomen has really expanded my horizons. I've enjoyed being able to help others, and have made many friends who I chat with on email and on Facebook as well.

HipWomen is a group of women who have hip problems, usually dysplasia (but not exclusively). Like you, they found they needed someone to talk to about it. It was started in 2003 by Sarah Jurgensen, a young woman with DDH. Since that time is has grown rapidly, and has just over 1,300 members at time of writing. Some people come and go, as their need for support waxes and wanes, while others stick around for years.

The HipWomen group is accessed via Yahoo, and is an email list. This means that you simply send an email to the group email account, and everyone in the group receives a copy of your message, and can reply to it. You can opt to receive individual emails, or a daily digest (which is a collection of all the messages sent in one day, put into one lengthy email), or receive no emails, and just read the messages on the web site. It is a *very* active list, with a *lot* of emails sent every day. You may like to set up a special HipWomen folder in your email program, and a filter to put all the emails into that folder.

I remember the first time I made contact with an online hip group, and found people who had been through the same issues as me — I cried and cried because before then, I had felt like the only one! I think that's when my diagnosis really hit home. **Freja, 45 UK**

To join HipWomen you need to be approved, which is part of the signing up procedure. Don't panic, there are no special tests required, this is just a way that the group moderators stop spammers from joining the group. Just tell them in one or two sentences why you want to join the group, and you should be approved quickly, usually within a day.

What you can do on the HipWomen site:

◊ **Messages** — Sending and responding to emails is the main activity of HipWomen. You can reply to an individual sender, or to the group as a whole. This means if you have something that you want to discuss or respond to with just one person, that you don't want the whole group seeing, you can. These options appear in the "To" field of your reply email (*Reply to Sender*, or *Reply to All*). You can send message via your reguar email account, or via the HipWomen web page.

◊ **Photographs and x-rays** — These can be posted so you can post your x-rays and photos of your scars, if you like, and view others.

◊ **Databases** — There are several of these with exercise tips, things to take to hospital, lists of surgeons members have seen, lists of members' blogs, and so on.

◊ **History** — All the messages sent to the group are archived on the HipWomen web site. You can search for what you are interested in, and see what has already been talked about. There are over 35,400 messages in the archives at the time of writing, so there's plenty of experiences and advice to read through right there!

◊ **Making friends** — As with any group, whether you're meeting in person or not, you will find some people who you have a close rapport with — maybe you're of a similar age, or both studying the same thing at university, or have other interests in common, or having surgery in the same week. These people can become more than just HipWomen acquaintances, but close friends.

◊ **Meeting up** — Very occasionally HipWomen who live in a similar area organise to get together at a local café or park, which is a bonus for those who can get to one of these irregular gatherings — often with nearly everyone on crutches! They are sometimes arranged by interested members who might be travelling to a central city (like London) for a check up with their surgeon, for example. These are not a frequent feature of the HipWomen group, but are a great treat when they can be pulled off.

HIP CHICKS

www.Hip Chicksunite.ning.com

The Hip Chicks forum (Hip Dysplasia and Impingement Support Group) was set up by Krystal Clausen in 2010. It is similar in some ways to Facebook, so you have your own page and details, you can add other Hip Chicks as Friends, post discussion topics on the Forum, write blog posts, share photos, join in the real time Chat Room, and send PMs (Private Messages). There is even a great range of fun Hip Chick merchandise if you really want to get decked out! There are different sub-groups within the site (such as UK Hip Chicks, and Resurfaced Chicks) so you can join these as well, if you like. There is a minimal annual fee to join this forum, to help cover hosting costs.

OTHER USEFUL SITES AND FORUMS FOR SUPPORT

Facebook

There are several groups on Facebook devoted to hip dysplasia:

◊ Adult Hip Dysplasia Group
   < www.facebook.com/group.php?gid=50910135532 >

◊ Steps Charity
   < www.facebook.com/pages/steps-Charity-Worldwide/1150753038 >

◊ Hip Chicks Unite
   < www.facebook.com/pages/Hip-Chicks-Unite-Hip-Dysplasia-Awareness/340869876522 >

◊ The International Hip Dysplasia Institute
   < www.facebook.com/IHDIOnline >

Happy Hips

< happyhips.webs.com >

This is a UK-based website and blog for teenagers with hip dysplasia, and their parents, run by young adults with dysplasia. It has forums, news, photos, and many features for members. Membership is free.

Hip Universe

< hipuniverse.homestead.com >

This group is focussed on people with a range of hip problems, including (but not exclusively) dysplasia, who are looking at PAO, resurfacing, or hip replacement surgery.

International Hip Dysplasia Institute (IHDI)

< www.hipdysplasia.org >

The IHDI was formed by a group of specialists in America, and is based at the Arnold Palmer Hospital for Children, in Orlando, Florida. They have a medical advisory board with hip specialists from across the world. Alongside proving information for adults, children and their families on hip dyplasia, the IHDI conducts research into the condition, seeking both to better understand the condition, and improve treatments. They are a charitable organisation, and rely on donations to carry out their research.

IHDI on Facebook

< www.facebook.com/IHDIOnline >

This Facebook group is for both adults with DDH, and parents of kids with the condition. If you have a medical question and you post it here, the IHDI medical board typically responds.

About Orthopedics

< orthopedics.about.com >

A very informative web site with weekly newsletters and helpful articles on a wide range of orthopaedic topics, written by American orthpaedic surgeon Dr Johnathan Cluett. While it doesn't often cover DDH specifically, the general information supplied is often quite pertinent to hip patients.

Medical information online

The world of the internet has made medical information more accessible to all, although there is a lot of rubbish out there too. How do you differentiate between what is good and what is a load of twaddle? And how do you even find information about adults, instead of babies and dogs?!

When I look up information on the internet, I look at a couple of reliable sources, typically I check < www.emedicine.com > and < www.nih.gov > or other well-established sources ,and then I look for "firsthand experience" sources, which is how I found HipWomen and the FAI group. **Brenda, 48 USA**

I have to say that for me, the research was very important. It was helpful to Google hip stuff on my own and discover links, as opposed to getting them from someone. Because of that research, by the time I met my surgeon, I was pretty sure that PAO would be the way to go for me, I was prepared and had a great positive attitude towards it.

I think this was a bit of a better experience than for many people who first heard the scary word "hip surgery" and that they might need it — then did their research, full of doubts and uncertainty. But it is true that many people start researching after they hear about their situation, not in anticipation of it. **Arpine, 28 USA**

SEARCH ENGINES

There is quite an art to using a search engine effectively. There are a few easy tricks that can help you to find better quality information on the web:

◊ Use a minus sign before a term you want to exclude, such as *dogs* or *paediatrics*. For example, your search terms might read: *hip dysplasia -dogs -paediatrics*.

◊ Use a plus sign + in front of a term to force it to appear in the search results. For example, your search terms might read: *hip dysplasia +surgeon*.

◊ Putting a phrase inside quotation marks makes a big difference too — this forces the search engine to return results that have those exact words in their content. For example, your search terms could read: *"hip surgeons in New York."*

Google has a helpful Cheat Sheet if you want to delve further into how to use a search engine like a true pro:

< www.googleguide.com/advanced_operators_reference.html >

GOOGLE SCHOLAR

< www.scholar.google.co.uk >   < scholar.google.com.au >
< www.scholar.google.com >

This is the scientific arm of Google. Searches on Google Scholar will generate more documents from medical or scientific journals or databases, whereas plain Google searches are more likely to identify patient information, surgeons' web sites, and general discussions on dysplasia.

Certainly, as a patient, you want that kind of information too — but by going through Google Scholar and the medical literature, you can identify surgeons who have experience in dysplasia (who have written papers on the subject), which can then help you search more efficiently on Google. For example, a lot of the surgeons who deal with DDH have their own websites, which are a great source of trustworthy information. Mr Johan Witt's web site is one such example: < www.hipjointsurgery.co.uk >.

PUBMED

< www.ncbi.nlm.nih.gov/pubmed >

PubMed is the homepage of the U.S. National Library of Medicine National Institutes of Health, which is a database of medical, biomedical, and life sciences journals.

Searching through PubMed is quite simple, just put the key words you are interested in into the search engine and hit 'Search', as with any search engine.

The benefit is only articles that appear in peer-reviewed journals appear (i.e. they have been scrutinised by other medical specialists before being accepted for publication).

PubMed includes content from some well recognised journals for orthopaedic research including: *Journal of Bone and Joint Surgery* (JBJS; British, European and American versions), *Acta Orthopaedica, International Orthopaedics, Clinical Orthopaedics and Related Research, British Medical Journal* (BMJ), *Journal of Paediatric Orthopaedics, Journal of Surgical Orthopaedic Advances, American Journal of Orthopaedics,* and *Hip International.*

You can find articles written by any of the surgeons in your area, which can help you identify surgeons who specialise in hip dysplasia. You can also go into the individual sites of the journals to search for relevant articles.

MEDLINE PLUS

< www.nlm.nih.gov/medlineplus/ >

Medline Plus is a more patient-focussed web site, also run by the National Library of Medicine in the USA. It has dictionaries, a medical encyclopaedia, information on clinical trials, and articles written in more "everyday" language. The site has no advertising, and is a trusted resource.

ABSTRACTS VS FULL TEXT

In general, the abstract (summary) of the article is free to access. Some of the full texts are also free, but in general you will have to pay a fee (which can be quite high) to access the full article. Don't worry though, the abstract generally provides enough information, with the conclusions of the paper presented in a very concise form.

The full text of a paper gives you the entire article including references, diagrams, tables, and highly detailed information. Trying to read the full text of a medical paper, and picking out the stuff that is relevant to you, is quite hard work. Unless you have medical or statistical training (or both, and sometimes not even then!), you may struggle to gain enough information to justify the cost of paying for the full text. You can also see if your local state, university, or hospital library has subscriptions to the journals you're interested in — if they do, access should be free or very inexpensive (possibly just photocopying fees).

SOME GOOD MEDICAL AND HIP WEBSITES

◊ **Discovery Fit   Health — Hip Dysplasia**
< health.howstuffworks.com/diseases-conditions/musculoskeletal/
hip-dysplasia4.htm >

◊ **Facts About Total Joints**
< www.newtotaljoints.info >

◊ **Johan Witt, Joint Orthopaedic and Trauma Surgeon**
< www.hipjointsurgery.co.uk >

◊ **Hip Baby**
< www.hip-baby.org > (for parents of babies with hip problems)

◊ **Hip and Pelvis Institute**
< www.hipandpelvis.com/patient_education >

◊ **Medscape**
< emedicine.medscape.com/orthopedic_surgery >

◊ **National Institutes of Health**
< www.nih.gov >

◊ **Orthopaedia**
< www.orthopaedia.com/display/Main/Hip+dysplasia >

◊ **The Hip and Knee Institute**
< www.hipsandknees.com >

BLOGS

A blog is short for we**b log**, and is basically an online diary you write, that your friends and others can comment on. You may also find reading blogs of other people's experiences useful, as we have previously mentioned. The HipWomen group has an online database of blogs that people are happy to share. You can also use an internet search engine, specifying "blogs" to search for those related to the subject you are interested in, e.g. PAO, hip replacement, dysplasia, and so on.

You may find that writing your own blog is a good way of dealing with things. Writing down your emotions and experiences can help you come to terms with them, and you never know, you may end up helping others through similar situations. We discuss this in more detail in Chapter 7.

Here are some of hip blogs to get you started. Many of the have links to other hip dysplasia blogs, so from these few starting points, you can go much further. If the blog belongs to someone who has contributed to this book, their name follows their blog address. Not all of these blogs are active (frequently or recently updated), but they all have good personal information and stories about DDH.

◊ **Arpao** < arpao.livejournal.com > (Arpine)

◊ **Happy Hips** < happyhips.webs.com > (Hannah) For teens

◊ **HipPAOer** < hippaoer.blogspot.com >

◊ **HipSk8** < hipsk8.blogspot.com >

◊ **Hipster Club** < www.hipsterclub.com > (Jodi)

◊ **My THR Journey** < deethr.blogspot.com > (Deirdre)

◊ **My UK Hip Story** < myukhipstory.blogspot.com >

◊ **Jejune's Place** < jejunesplace.blogspot.com > (Denise) (search for "hip" to bring up relevant posts)

◊ **One Hip Chick** < inamarieandherhip.blogspot.com > (Ina)

◊ **Sophie's Hip Story** < sophie-hip-dysplasia.blogspot.com > (Sophie)

◊ **Wibbly Wobbly's Blog** < annickhollins.wordpress.com > (Annick)

How to spot pseudo–science sites

There are millions of medical web sites out there, with more appearing every day. Many medical sites provide good, evidence-based medical information and advice. However, there are just as many — if not more — which can lead you astray.

Before you spend your hard earned cash on that magnetic mattress or expensive supplements, follow our guidelines for spotting "quacky" web sites, which will only confuse you with dubious or misleading information.

First of all, check whether the device, supplement, treatment, institute, or doctor's name is mentioned on Quackwatch < www.quackwatch.com >. This site was established in 1997 by Dr Stephen Barrett, and is the top source of critical analysis of all things medical.

Next, assess the advertisement or web site against these criteria:

◊ **1.** The site uses anecdotes, testimonials, or celebrity endorsements to promote its products. Remember the maxim *"The plural of anecdote is not data."* Just because there are a lot of stories, doesn't mean it's proof. A product that relies heavily on anecdotes and testimonials is probably unsupported by experimental evidence.

◊ **2.** The site warns you not to trust your doctor, and claims that "the authorities" are suppressing their information. Any "What *they* don't want you to know" sort of line is always a red flag.

◊ **3.** A product or treatment that is based on ancient wisdom is always suspect — it's a sure thing that it's pseudo-science. The nature of good science and medicine is that new theories are constantly being put to the test, and improved and changed as our understanding of any subject grows. Surgical and medical treatments are always being improved. An adherence to centuries-old knowledge, that hasn't changed in all that time, isn't really something to be proud of!

◊ **4.** If a treatment or product is first announced through mass media, rather than peer reviewed journals, it's another warning sign. Good science and medicine is published in science or medical journals first of all, reviewed, discussed, and repeated by other scientists and doctors, and then *eventually* becomes a tested treatment, medication, or product for patients.

◊ **5.** Any claim based on existence of energy fields or life force energy is suspect. No research — and there has been plenty — has *ever* revealed the existence of such forces. Medical treatments or healing treatments based around these sorts of energy fields are bogus.

◊ **6.** Does the claim sound too good to be true? It usually is. Is the deal too good to be true? It usually is. Does it claim to treat a vast range of different conditions? Another warning sign.

◊ **7.** Do the doctors have legitimate credentials? Check whether their Alma Maters actually exist, or their degrees have been obtained through recognised accreditation institutions. Unscrupulous people *do* get those free PhDs, and make use of them.

◊ **8.** "All natural" doesn't mean it's safe or healthy (poisonous toadstools, lead, and salmonella are all natural, after all!). Medications — while they do have side effects — have to go through incredibly rigorous safety testing procedures before they're allowed out in public. Natural remedies don't necessarily have such testing, may not have their side effects listed anywhere, and may not even be that "pure", but could be contaminated with other substances. There are many cases of Chinese medicine being adulterated with unlisted medicines and other undisclosed active ingredients, for example.

FURTHER READING

These books (and many others in the same genre) can help you to distinguish the "woo" from the evidence-based medicine.

◊ *The Demon-Haunted World* by Carl Sagan

◊ *Trick or Treatment* by Simon Singh and Edzard Ernst

Charities

STEPS

< www.steps-charity.org.uk >

STEPS is a small national charity based in the UK that helps children and adults with lower limb conditions, including developmental dysplasia of the hips. Their motto is "We don't take walking for granted", which is very appropriate, as you may well appreciate! It can put you in touch with other people in your situation, provide information, and has online forums where you can discuss your condition, ask questions, and generally find support. They also have a page on Facebook which you can follow.

THE INTERNATIONAL HIP DYSPLASIA INSTITUTE

< www.hipdysplasia.org >

The International Hip Dysplasia Institute (IDHI) is a not-for-profit organisation and research group, working to improve the health and quality of life of those afflicted with hip dysplasia. It provides up to date information for both adults and children, and produces information leaflets for medical professionals and patient. They also have an active Facebook page < www.facebook.com/IHDIOnline >. They rely entirely on charitable donations to perform research and provide education.

Finding a surgeon online

The process of being diagnosed with hip dysplasia means you must have seen someone who at least has the skills to recognise the condition.
In general, your family doctor will give you a referral to a specialist. Sometimes that specialist won't be able to help, and will refer you on to yet *another* specialist. The process can take some time. We discuss this process in more detail in Chapter 10.

Sometimes your physiotherapist, or other allied care professional, will be the one who encourages you to look for more significant treatment and a specialist. They cannot provide you with a referral, but they can often suggest knowledgeable surgeons.

If you have not been referred onto a specialist, here are some tips that can help you identify a good surgeon:

◊ Research medical papers for surgeons who have an interest in hip dysplasia (the Cited References at the back of this book are a good place to start!). HipWomen and other online forums are a great resource and can help you identify surgeons in your area.

◊ General orthopaedic surgeons, if they have the ability to diagnose hip dysplasia but not to treat it, can often recommend another surgeon you can see.

◊ The Pediatric Orthopedic Society of North America (POSNA) can advice on paediatric surgeons in your area for those living in the USA. < www.posna.org >

◊ For those in the UK and Australia, you may find you have to go through your GP to get a referral, they can recommend a specialist to see. You may need to see a general orthopaedic consultant first, as specialists in hip dysplasia sometimes require a referral from another specialist instead of a GP.

◊ Internet search engines, see our tips above.

The main thing to remember with finding medical information online is that you need to be careful of the sources of information you read. But there are *many* people out there (out of the world's 7 billion inhabitants, roughly 0.13% have hip dysplasia, which is around 9 million people) who are going through the same thing as you. Even if you never meet in person, they can be a great support for you, and you can support them too!

Chapter 18
For Carers

This chapter is not aimed primarily at the patient (although they're encouraged to read it too, as the "caring relationship" is two-way!), but for you. You, who supports your loved one through all the hard times associated with hip dysplasia. Through years of increasing disability before possible surgery, the time of surgery, and the long period of recovery after surgery. And there may well not be just one surgery to get through, but many. The role of carer is generally a poorly appreciated one. You're putting everything into looking after your "hippie" (by which we mean hip patient, not a flower child of the 1970s, although they may be one and the same!), usually with very little support or recognition of how hard it is for *you*.

First of all, we would like to say that none up of us hippies manage without our carers, and all their hard work and support. Even if society at large doesn't appreciate what you do, your hippie certainly does! Your help is invaluable and essential, and we take our hats off to you. We hope this chapter offers a bit more insight into the particular challenges that a hip dysplasia patient presents, and ways that you can look after yourself better. You can't care for someone else if you don't look after *yourself* first of all — if you're not functioning well, nothing else is going to work!

Words in bold can be found in the Glossary.

What caring involves

The role of carer involves wearing a great many hats — you may need to be a housecleaner, cook, chauffeur, sole parent, personal shopper, and gardener, as well as a personal nurse, advocate, counsellor, and personal trainer at times! Your hippie will also have times when they need a lot of help, such as immediately after surgery, and times when they just need a little help now and then (such as a bit of help getting up from sitting on the ground). Hopefully most of the time your caring duties won't be too onerous — and most patients are keen to maintain as much independence as possible, for as long as possible. The time when you will be most in demand is during the recovery period immediately after surgery.

Your hippie may appreciate it if you can accompany them to their various medical appointments, especially with specialists, both before and after surgery, so that you can write notes, ask questions, and be aware of all the intricacies involved in their case. Ask your hippie if this is something that they'd find helpful.

At times you may need to do almost everything for your patient — from helping them shower, to helping them dress, and get in and out of bed. There can be a fair amount of lifting involved, and it's important to look after your back and lift safely. There is information on how to move a patient from a wheelchair, or in and out of bed and chairs and so on on the internet, but we list a few basic tips and tricks here below.

Surgeries for hip dysplasia

Here is a brief guide to the different surgeries for hip dysplasia (also called DDH, for **developmental dysplasia of the hip**) and their recovery times, so you know what to expect (there is more detail on these in Chapter 4). In general, most hip surgery patients will be back to their normal activities within three to five months, but complete recovery can take a year.

OSTEOTOMIES

An **osteotomy** is any surgery that cuts through bone, to shorten it, lengthen it, or change its alignment.

Periacetabuar Osteotomy (PAO)

The most common osteotomy for hip dysplasia is the **periacetabular osteotomy**. In this major surgery, the surgeon cuts through the pelvis in several places, effectively freeing the hip socket from the pelvis, and realigns it to help create a more normal hip socket. A PAO can only be carried out on a relatively arthritis-free hip, so if your hippie is having a PAO, they probably don't have much arthritis in the joint, or too much hip pain.

The bones are repositioned, and typically held in place with long screws (which are usually eventually removed). Your hippie will be **non-weight-bearing** (not allowed to stand or walk with full body weight supported by their feet) on crutches for six to 12 weeks, and recovery is typically quite slow. There are some restrictions to movement that the patient has to adhere to post-op (more on this in Chapter 15).

Femoral Osteotomy

Less common is a **femoral osteotomy**, which is used to correct deformities of the femoral head like **coxa valga** and **coxa vara**. The patient has to be non-weight-bearing on crutches for six to eight weeks, and again there is a very long recovery period.

Total Hip Replacement (THR)

This surgery is also sometimes called a **full hip replacement (FHR)**. The old deformed arthritic hip joint bones, comprising of both the hip socket and the head of the femur, are cut off and replaced with an artificial hip joint (socket and ball). The good thing about this surgery is that the pain from **osteoarthritis**, that has probably been making life a misery for your hippie, will be gone instantly.

Recovery from the surgery itself is another matter, and can take quite a few months. Your hippie will probably have very strict hip restrictions for at least three months (see Chapter 15 for details). They will be encouraged to weight-bear right away (ie leaning with their weight on the operated-on leg), but will be on crutches and/or a walking stick for some weeks.

Resurfacing

A **resurfacing** is when the hip socket is lined with an artificial hip socket. The head of the femur is *not* removed, but simply covered with a smooth cap. There is much less bone removal than with a total hip replacement, and revision surgery (replacement) is just easier overall. Not all patients are candidates for this surgery, but is very successful for those who have it. Resurfacing has similar recovery times to THR, although there may not be such strict hip restrictions, if any.

Revision THR

A **revision** hip replacement is a difficult surgery. It is done when an existing hip replacement has worn out or has some other major problem, and needs to be replaced (or "revised"). It is a more complicated surgery than the original hip replacement, as the old prosthesis needs to be cut out of the bones, and a new one put in, with less bone for the surgeon to work with in general. It can have a longer recovery period than for a THR, and there is also a higher chance of complications with a revision.

A balancing act

It can be tricky gauging the different levels of support your hippie needs. One day he or she might be in a terrible amount of pain and unable to function well. You would need to step up the level of care you're providing, waiting on them hand and foot, attentive to their every need. The next day, they might be feeling much better, but if you still try to provide that same "high level" of support, they might get annoyed and say they're perfectly capable of doing that for *themselves, thankyouverymuch*! They're not trying to be difficult and make your life hard, it's just a very tough time.

This sort of constantly changing need for care can make for a difficult balancing act, and a tense relationship at times. It's best if you and your hippie can come up with some way of easily communicating to each other what sort of level of care they feel they needs that day, or some signposts for what symptoms or problems require a change in level of care. Perhaps you can come up with a code system, such as "I'm having Code Red Day today" (indicating a need for a lot of care), or "Today's a Code Green Day" (I'm coping pretty well alone).

Let your caregiver know what the thresholds are for getting more help, such as signs of infection, and milestones to watch for in your decline or improvement. And let them know what the thresholds are for taking a step back, i.e. your personal space and theirs, care fatigue (for them and you). Identify milestones in your improvement that will allow you to do more for yourself. **Brenda, 48 USA**

Some practical tips

WHEELCHAIRS

Not all hip patients end up needing a wheelchair, but it can be a very useful way of getting around in a shopping mall, museum, or other large venues, or during travel. If your hippie is in a wheelchair, learn how to use it safely from the occupational therapist at the hospital, or their physiotherapist or doctor.

Here are a Few General Tips:

◊ Make sure the patient keeps their hands and arms on their lap, to avoid accidents with being caught up in the spokes. Keep an eye on their feet as well!

◊ When going down a ramp, walk down backwards, pulling the wheelchair backwards behind you. This will give you better control. It can be a good idea to have someone else help, by holding onto the front of the wheelchair, when going down ramps, for extra safety.

◊ When going up a ramp, it can help to have another person help by pulling the wheelchair from the front while you push from behind.

MOVING THE LEG

Be very careful when helping your partner move their operated-on leg, especially in the first few weeks. It can be excruciatingly painful if it's moved in the wrong way, or even just too quickly or suddenly. Your hippie's nurses, occupational therapist, or physiotherapist should provide information on how to safely move the affected limb, but if in doubt, just check with the patient as to what hurts and what doesn't.

Carers should let the patient tell them how they need to be helped to move their leg, not just assume. I say this because of my first experience, and reading about others where their carer has just lifted their leg onto the bed too quickly, which has caused a lot of pain for the patient. **Lea-Anne, 40 Australia**

363

STANDING UP

Here are a few tips on how to help your hippie stand up, without hurting your back.

◊ If your hippie isn't able to weight-bear at all, or weighs more than you do, you will need a third person to help you move them safely.

◊ Make sure your hippie is wearing non-slip shoes or slippers.

◊ Ask your hippie to sit as close to the edge of the bed or chair as they can, with feet apart (roughly hip width apart).

◊ Put your arms around their chest, and clasp your hands behind their back.

Figure 18.1: How to help a hip patient stand up safely.

◊ Put one of their legs between your legs.

◊ Rock back and forth slightly to get a bit of momentum going. On the count of three, raise your hippie to a standing position, and they can push up too with their arms if there are armrests to push against.

◊ Once they're standing securely, hand them their crutches or walking stick.

If you then want to move them to a chair or wheelchair, this is what to do next :

◊ Place the edge of the chair or wheelchair gently against the back of their legs. If you're using a wheelchair, make sure the wheels are locked in position with the brakes.

◊ Keep your hands firmly clasped behind their back, and slowly lower them into the chair. You may need a third person to hold the chair steady while you do this.

For more information, including diagrams, on how to move a patient safely in different situations, search the internet with such search terms as *caregiver, carer, safety, lifting, moving, patients.*

Never let anyone help you to stand up by pulling you up to your feet by your arms, as they are forcing you to put all weight onto your hips. I still have people putting both hands out to me offering me help getting up, and have to be very tactful letting them know that it wouldn't be any help, and might harm me.
Margaret, 60 UK

PREPARING FOR SURGERY

It's important to help set up a safe home environment before your hippie goes into hospital. Stairs will be a problem at first, so your patient might need a bed set up downstairs for a week or so. Other things to look out for are tripping hazards (electric cords and slippery mats especially). They won't be able to lift their feet very high for a while, so stepping over things will be a problem, and they're more likely to trip, stumble, and even fall. This puts their hip at risk of dislocation (especially after a total hip replacement), apart from general bruising and fright.

There are a bunch of useful disability aids that make life much easier both before and after hip surgery. Things like long-handled grabbers, sock gutters (for putting socks on), long shoe horns, and so on help your hippie be more independent. It's also important to start lining up friends, neighbours, and family to help, especially immediately after surgery. If a lot of people can each offer a little help, it can make a huge difference to you. The occasional help with providing transport for your hippie to a medical appointment, a meal now and then, or picking the kids up from school can give you a break from the demands of being the sole "fully functioning" adult in the household ... We cover all of this in detail in Chapter 11.

It's important to remember that before surgery your hippie may be rather snappy or short tempered, especially if he or she is in a lot of pain all the time. The pain and frustration can get to be too much quite quickly. They may be teary, scared, angry, frustrated, grieving, and generally difficult to live with.

There will be a long time — starting before surgery and lasting at least six months afterwards — where everything will seem to be about her. That's because it is. It's the same with any kind of surgery. You become very inward-focused because the pain, the upcoming operation, and then the recovery afterwards — those hips and the surgery aren't just the most important thing in your life —they *are* your life at the moment. **John** (carer)**, 39 UK**

It is worth remembering that no matter how upbeat you are about it (which is great), she will undoubtedly be feeling some (and possibly quite a lot of) fear before her surgery. This may make her a bit introspective, grumpy and/or self-centred — she can't help it — it's the body's fight or flight mechanism kicking in. So don't be surprised if she is a bit distant or moody. **Freja, 45 UK**

THE FIRST WEEKS AFTER SURGERY

It's a good idea to take a week or two off work around the time of your hippie's surgery, if at all possible. You will be extra tired, running around, and it's best not to be distracted by work and other demands.

The most intense work you'll have to do in caring for your patient is in these first few weeks after surgery. Even though it is terribly hard initially, you should see some small level of improvement in their condition, and level of independence, every day.

If you notice that their condition is actually deteriorating, with an increase in pain or fever, increased redness or weeping of the wound, or a reduction in how much they can move or do, *don't ignore your observations*. If they have an infection or an illness, it's vital that it's caught as quickly as possible. Your close daily observations of their condition can be crucial.

 If they are getting sicker in any way, be proactive, get them to the doctor, or even call their surgeon if you suspect an infection (there is more information on the symptoms to look out for in Chapters 4 and 14). Trust your instincts on this one! It really is a situation of better safe than sorry.

The anaesthetics given during surgery are very strong, and take several days for them to get completely out of the body. The physical trauma of surgery has an even bigger impact, more than you may realise. When you add in the heavy duty pain medications your hippie will be on in the first weeks post-op, you can see that you may have a rather dopey, listless, and befuddled patient on your hands. Be aware of this, and make allowances. They won't be their normal selves.

Become aware of exercises your partner needs to do, what medications they need to take, wound care, nurse or doctor visits, physiotherapy appointments, and so on. Encourage your hippie to do their physio exercises regularly.

Showering and toileting can be difficult initially, especially with stepping into and out of the shower recess. If your only shower is in the bath, your hippie may need sponge baths for the first little while, until they're more secure and able to step into the bath. There is more information on bathing post-op in Chapter 14.

Your hippie may need help with putting on anti-embolism **deep vein thrombosis (DVT)** stockings, shoes, socks, and getting dressed in general.

Your patient will be unable (and in fact, not allowed) to drive for quite a while after their hip surgery, so you will need to be their chauffeur for some time, most probably a couple of months. Initially they won't feel up to going out much at all, but they're likely to have fairly frequent medical and physiotherapy appointments. They may need to get out to have blood tests done, if they're on certain medications like warfarin. After a few weeks, going out on a short shopping trip or to visit a friend may seem more appealing and feasible.

One important thing a carer can do is to find a way for the patient to continue to get out and do things. The more you sit at home, the more time you have to feel sorry for yourself, which is really easy to do. I had my first PAO just before Christmas, so my husband hired a wheelchair (just one from the mobility shop at the local shopping centre) and wheeled me around the so I could still do my Christmas shopping. It was good to just get out and about and feel normal again. **Sandra, 39 Australia**

You may have to do things that you're not used to, like making school lunches for the kids, doing the grocery shopping, doing the laundry, cooking and housework, driving the kids to sporting activities, or running errands.

After a hip replacement, and some other hip surgeries, your patient will not be allowed to do twisting actions, so chores like vacuuming, raking, and sweeping are absolutely *not allowed* — even if your hippie physically feels up to it, their hip joint is too weak, and has a high risk of dislocating. The ligaments and muscles that hold the joint capsule together have been cut, and they take weeks to recover. It's not worth the risk, just for a clean floor!

It is the silly things that I can't do, like getting the wok out of the cupboard because it is too heavy, or carrying the dogs food bowls when full. Everything just takes so much longer to do, but my husband has been around, and my kids, and they have just had to help out a lot more — getting the washing done and hung out, hoovering, running back upstairs because I forgot my book.
Annick, 47 UK

Don't be surprised if your hippie cries easily and frequently after surgery, for some weeks. It's no reflection on how well you're supporting them, but just from the pain, stresses, and frustrations of recovery from major surgery. It's a very common reaction. We write more about this in Chapter 14.

Throughout all of the run-up to my wife's resurfacing, and in the post-op period afterwards, I was never in any doubt that she was grateful for everything I did for her, but I also realized that I had to let things run to her schedule — her emotional schedule more than anything else. If she wanted sympathy and time to vent, I had to allow her that, just as when she needed support and upbeat positivity from me, I had to give that, too. I also had to recognize that there would be a time when she wasn't going to be quite so inward-focused and just to wait for that time to come along naturally. One-word summary? Patience.
John (carer)**, 39 UK**

ASKING FOR HELP

It can be hard to ask for help, both if you're not quite handling things at home and *you* need help, and for your hippie to ask you for help. Even if you *are* handling things well, it's good and healthy for you to have a break, even just a few hours out with friends, or whatever you find relaxing.

There's no shame in asking for help if you're really struggling with meals, housework, or the kids, for example. This is where extended family and friends can really come to the fore, and most people are only too happy to help out if asked to do a particular task. It's better to ask them for help with a specific thing, like can they pick the kids up from school on Wednesdays, or cook a meal for the weekend, rather than to "generally" ask for help.

SEX LIFE

If your patient is your partner, your sex life probably hasn't been so good with their increasing disability and hip pain, and more stress on you as carer. It won't be good for some time after surgery either, but the good news about hip dysplasia is that once the surgery has been done, and your partner has recovered from it, they will be *much* better, and life should get back to normal within a few months.

At least you can be fairly sure that after all the hip surgery is over, you and your partner should be able to re-establish an enjoyable sex life without pain. You will always need to be a bit careful of The Hip, but so long as your partner adheres to their hip restrictions, your sex life should be much better than before! We discuss sex and hips in more detail in Chapter 16.

OUTBURSTS AND FRUSTRATIONS

The time leading up to major hip surgery can be very frightening and painful for hippies. The surgery itself is pretty brutal, and recovery can be slow, with setbacks and complications sometimes making life harder. Even the smoothest of surgeries with a "textbook" easy recovery is still difficult, painful, and takes a lot of time.

Although, hopefully, most of the time your hippie copes fairly well, sometimes it may all be too much, and they may need to rant, complain, and carry on a bit about being in pain, being fed up, frustrated, angry, scared, or whatever. At times like this, they may just need you to hear

369

them. It may be that all they need from you in response is to hear: "I'm sorry you're having a bad day, I know how hard it is for you."

You can even ask your hippie *exactly* what it is they want to hear from you when they're ranting, and you just repeat as directed when the time comes! This ploy is surprisingly effective. They aren't expecting you to come up with solutions for their problems (unless they specifically ask for that), but simply for you to listen to them, and acknowledge their pain and suffering.

GETTING AWAY FROM IT ALL

It's important for you both to have a break from each other, and especially for you to have a break from your caring duties. Even if it's just going down the street for a coffee in a café by yourself, out for a beer with friends, a drive in the country, or a trip to the swimming pool. It can make all the difference if you're feeling wrung out, exhausted, annoyed, and trapped at home.

At least with hip surgery your partner shouldn't be *totally* dependant on you 24 hours a day (although it may feel like that initially!). It should be possible to arrange to get out by yourself for an hour or two, leaving them set up in a comfortable chair or bed, with a phone by their side in case of emergencies. Make sure the remote controls for the TV, DVD player, and stereo are by their side, too! Make sure they have a drink by their side, and are able to get to the toilet without you around (maybe another helper can come over while you're out?).

Don't forget that sooner or later it will do you both good if you do something separate from each other — for example, if an old friend calls, maybe you should go out for coffee. When my husband and I did this, it worked really well — it made me feel more independent, and gave him a break from thinking about me all the time! **Freja, 45 UK**

Comments from hippies

INDEPENDENCE

If I feel that I can do it, let me — if I can't, I will ask for help. I'm not a person who can just sit and put my feet up, and be waited on hand and foot — it would drive me nuts, I have to be able to participate. My family let me get on with things, but knows that I will need a hand at some point. **Annick, 47 UK**

I think the experience of being cared for is a challenging one, especially if you are an independent person not used to asking for help. I really struggled, especially by my last operation (THR) to ask for help, as I was fed up with having to rely on others — although in reality, it was probably more of an issue for me than for others. I must admit that one of the things that puts me off having a PAO on my other hip is going back to being reliant on someone else again for weeks and weeks. **Deirdre, 36 UK**

I think it is good to have a carer who knows you very well, and is willing to do pretty much everything for the first four weeks at least. But they must not wrap you in cotton wool, they have to let you be in charge of your own body, and know what you are ready to do. An example: I wanted to drive, but my mother-in-law was too nervous, and wouldn't let me until I got the doctors go ahead. But the fact that I wanted to try shows that I was ready and needed to gain some independence and control over my own environment again. I ended up taking the car around the block on my own when she wasn't around, so I could prove to myself and her that I could do it! **Lea-Anne, 40 Australia**

You can probably manage on your own during the day if you organized yourself — like having a flask of hot water, phones nearby and so on. I was actually okay with going up and downstairs right from the word go after my THR, but I remember after one operation a couple of years ago, I didn't brave the stairs more than once a day — would get myself downstairs and stay there all day (which is okay if you have a downstairs toilet!). **Deirdre, 36 UK**

My mother-in-law came up to stay with us for about a week after my THR. My husband didn't really want anyone coming in, but as he has chronic illness that makes him very fatigued all the time, I insisted. I didn't want to have to put so much pressure on him, and knew that his mother would be a great practical support around the house, which she was. Having her there freed my husband to concentrate on the more immediate and intimate things I needed, like help with showering and dressing. **Denise, 47 Australia**

EMOTIONS AND PRIVACY

A nice thing to do is to bring your hippie in a little present — not right after surgery, but maybe a day or two after. My husband brought me in a custom made T-shirt (the t-shirt said "Freja 2.0") . It just made me feel that he still saw me as a person, not just a hospital patient. It was good that he didn't bring me flowers or other traditional hospital gifts, he brought me something fun instead. **Freja, 45 UK**

Be aware of the mental toll this hip condition can have on the patient, both before surgery and post-operatively. It can be quite frustrating and even depressing at times, when you are in constant pain and not able to do the things you were once able to do easily. This is especially hard if the patient is young and/or had an active lifestyle.

It's something you are reminded of with every step you take, so there's no escaping it, it's on your mind constantly, and can easily become all-consuming. And it can be equally frustrating and depressing post-operatively — the long and slow recovery, the difficulty sleeping, the inability to care for yourself and your family, and the restrictions and limitations of the walking aids.

When you aren't able to do every-day tasks like put on your own socks, get dressed standing up, carry a cup of coffee to the table, or bathe your children, it can be incredibly frustrating. It is important that a carer understands this, not just to offer practical support, but to monitor your mental state and general wellbeing. **Sandra, 39 Australia**

Remember that despite the fact she may need some intimate stuff doing for her (for example I needed help putting my underwear and socks on, and needed help getting into and out of the shower) she will still need a little privacy. Getting someone to help putting your knickers on is lovely and helpful, but can make you feel like either an ill old lady or a helpless child — it does not make you feel sexy or in control of things! **Freja, 45 UK**

PRACTICALITIES AND PREPARATION

Establish a routine and try to include the caregiver before you are incapacitated so the processes can be shared and run more smoothly for all. **Brenda, 48 USA**

When I was left at home on my own, my husband made me lunch and cut it and put it on a plate covered with Gladwrap, so I could carry it safely while on crutches. That was very helpful! **Lea-Anne, 40 Australia**

I did as much as I could in advance — made meals to stock up the freezer, and organised the house to make things easier, but you do still need help, especially in the first two weeks after any major surgery. It's unrealistic to think you don't. Donations of freezer meals from family and friends were also very welcome! **Deirdre, 36 UK**

Use a notebook or journal to record information at appointments, questions that arise, observations — carry this to every appointment and share it with the caregiver. As you write during the appointment, don't be afraid to ask the physician or physical therapist to repeat something for you. You might also show the physician what you've written and ask, "Is this correct?" to ensure that you capture the information correctly. **Brenda, 48 USA**

I'm very lucky that my husband is very hand-on, and actually was happy to take on some household chores like hoovering which I struggled with pre-op anyway), and cleaning/washing floors (even better, he still does them a year on!). My mum lives just round the corner, so she would come round to make lunch and check I was okay each day for the first couple of weeks after my hubby went back to work. **Deirdre, 36 UK**

I paid someone to walk my dog every day at lunchtime, as I couldn't get out with him initially, and I felt better about that than having to rely on others. **Deirdre, 36 UK**

Establish a place for meds and supplies that you need; labels will help others to know what you have and where it belongs. Set up a schedule for things that need done at specific times, especially meds. **Brenda, 48 USA**

Suggestions for the patient

Remember to be a *patient* patient — your carer isn't there to be a complete slave to your every whim and desire. They're doing a tough job, running the family, possibly effectively being a single parent, doing unfamiliar tasks, not getting enough sleep, trying to keep up with their *paid* job, and worrying about you.

Try your best to not be too demanding, make allowances, don't expect the housework to be done perfectly, and be grateful for whatever meals are offered to you, even if this means frequent cheese on toast, tinned soup, or take-aways. A change in diet for a few weeks won't cause any lasting damage. The post-surgery weeks are a big challenge for them, as well as for you.

Let your caregiver know what your top three to five priorities are, and how you prefer that the priorities are handled. Privacy, pain management, and safety are some things to consider. **Brenda, 48 USA**

It's a nice gesture to organise some thank you gifts for your main carers before you go in for surgery. Whether it's a gift voucher, something special purchased online (wonderful if you can't get out much because of your hip), a handmade item, or the promise of dinner out at a favourite restaurant, they will really appreciate it.

If you can, prepare ahead of time a token of appreciation for your caregiver — a card, a small gift, a gift card for a cup of coffee, or whatever would be a nice treat for them. Encourage your caregiver to take some time for themselves.
Brenda, 48 USA

We hope that you (the carer) has found this chapter helful, and that other sections of the book can answer any questions you might have. And we hope that you (the hippie) has found some useful information here, on how to be an even better patient!

Chapter 19
DIY Projects

In this chapter we present some very practical things that you can make to help with your hip journey. You don't *have* to make anything, of course — most of these items are available commercially. But they're all seated activities, rather fun to make, and can help you feel more empowered on the path of your hip journey.

Bed ladder

If you're handy with wood, this is a simple project, and is a very useful disability aid for the first week or two after getting home from hospital. Note: this ladder will only work on the sort of bed where you can tie something securely to the foot of the bed, so either sturdy legs, or a bed frame that can take some stress from pulling.

MATERIALS & EQUIPMENT

◊ 2.5 cm/1" wooden dowelling, approximately 1.0–1.5 metres (3'– 4')

◊ Sturdy rope or cord, roughly 4 times the desired length of the bed ladder

◊ Electric drill (drill press by preference)

◊ Drill bit that is wide enough to allow the rope pass through the resulting hole

◊ Hand saw

◊ Ruler

◊ Pencil

METHOD

◊ **1.** Decide on the length of rungs you'd like (wide enough for you to have both hands on a rung, and some extra at the sides is best), roughly 25-30 cm/10"-12" is about right.

◊ **2.** Measure and mark this length on the doweling. You only need 3 or 4 rungs.

◊ **3.** Cut the dowelling into pieces along these marks using the hand saw.

◊ **4.** Mark a position about 3 cm/ 1" in from each end of each piece of dowelling. These marks should be parallel. See Figure 19.1 below.

Figure 19.1 The positions for the holes for the bed ladder rungs

◊ **5.** Drill holes at these marks, for the rope to pass through. These holes need to be parallel, and are best done with a drill press (easier, more accurate, and safer). Clamp the rung and drill the holes.

◊ **6.** Thread each end of the rope through the two holes on the bottom rung, so there are even amounts of rope on both sides. Then tie a knot in each side where you want the next rung to be, slide the next rung on and repeat until finished. The knots need to sit on the side you will pull against.

Make sure you have enough rope at the end to tie on to the end of the bed frame or legs. Tying the knots uses more rope than you might think.

See Figure 11.6 for an illustration of a bed ladder in use.

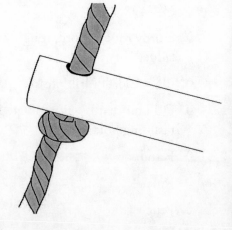

Figure 19.2 The knots sit below the rungs

HOW TO USE

Tie the loose ends of the rope ladder firmly to the foot of your bed frame. Lay the ladder on top of the bed. To sit upright, simply "climb" up the ladder rungs, pulling yourself upright as you go.

You may like to fasten a long rope handle onto the side of the ladder, and let the ladder dangle down the side of the bed; you just pull it up with the rope handle when you need to use it.

Baking soda poultice

These instructions are courtesy of Dr Rob Reid, sports physician.

A baking soda poultice is recommended for swollen and inflamed joints. It can be quite soothing on things like sore knees (which can occur with hip problems). It acts by decreasing the swelling in and around a joint, and draws out fluid (**oedema**) from under the skin. The joint swelling may be because there is fluid within the joint, or that the covering of the joint (the synovium that holds the fluid in the joint) is "waterlogged" and this produces a "boggy swelling" of the joint.

The bicarbonate of soda poultice draws the fluids out of the joint over a period of time. This generally takes seven to 10 days, especially for a joint that has had a boggy swelling for more than one or two days. The poultice can be used in association with any other treatment, and does not cause any skin problems.

MATERIALS & EQUIPMENT

◊ 2 cups of baking soda

◊ A piece of scrap fabric (old t-shirt etc), roughly 30 cm/12" square

◊ Some water (warm or cold, up to you)

◊ Mixing spoon

◊ Mixing bowl

METHOD

◊ **1.** Put the baking soda into the bowl. Mix in enough water to make a firm paste, roughly the thickness of bread dough. It should be soft and moist, but not runny.

◊ **2.** Place the lump of baking soda in the middle of your square of fabric.

◊ **3.** Tie up opposite corners of the fabric square, to make a neat little bundle. There should be just one layer of fabric under the poultice. See Figure 19.3.

HOW TO USE

Sit or lie down for an hour, with the poultice sitting or lightly strapped onto your swollen knee, ankle, wherever ... keep a face washer or towel handy, as it can drip a bit.

After the hour is up, put the baking soda mix into a lidded container, rinse and dry the fabric. You can reuse the baking soda repeatedly (just add a bit more water to get the paste consistency again).

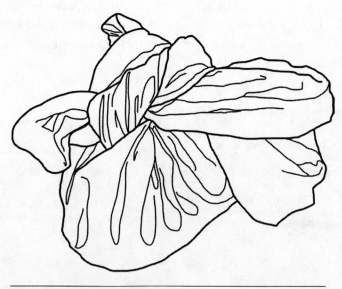

Figure 19.3: Baking soda poultice

Dry shampoo

Dry shampoo can be useful at hospital or home if you're not able to have a shower for a few days, and are suffering from a bad hair day. You *can* buy this, possibly from a hairdresser's or a pharmacy (Boots in the UK routinely stocks it). If it's hard to find, or you prefer to make your own, this easy recipe will help!

MATERIALS & EQUIPMENT

◊ Some powder, such as: baking soda, cornstarch, corn meal, semolina flour, wholemeal flour, orris root powder, ground oatmeal, and/or baby talcum powder. You only need ¼ cup or less.

◊ Shaker bottle with a lid

◊ A few drops of perfume or essential oil (optional)

METHOD

Put the powder, or a mix of powders, in the shaker bottle. If you wish to add some perfume to your powder, spray a little into the powder and mix carefully first.

You may like to experiment with different combinations of these ingredients to find your favourite blend (e.g. oatmeal + baking soda). If you have dark hair, try adding a bit of cocoa powder to the mix!

HOW TO USE

Sprinkle a little of the powder into your hair. This is a messy procedure, so best done with a towel over your shoulders and lap, or leaning over a sink. Short hair only needs a few teaspoons of powder, long hair might need 2 or 3 Tablespoons. A little goes a long way!

Use your fingers to massage it around your hair and scalp, and comb it through your hair. Let it sit for about 5 minutes. Then brush your hair thoroughly to remove the powder. The powder will absorb the dirt and oil from your hair, no need for water! It won't be as good as a regular wash in the shower, of course, but it's a reasonable substitute if you're not able to shower for a few days.

Remember to clean your brush or comb afterwards. If your fair gets frizzy after this treatment, brush it with a wet brush or comb. Baking soda, in particular, can be very drying, so use it sparingly.

Heat pack

Heat packs are a well-known and widely used remedy for sore and aching joints and muscles. They are very easy to make at home.

MATERIALS & EQUIPMENT

◊ Small piece of fabric, 100 % cotton by preference, or another natural fibre, or a cotton polyester blend. The fabric needs to be able to handle being heated in the microwave.

◊ A few cups of filling: dry rice, wholegrain wheat, buckwheat, millet, barley, flax seeds, lentils, etc.

◊ Sewing machine, or sewing needle and thread

◊ Scissors

◊ Pins

Optional

◊ Scented additions (optional): cloves, dried lavender, nutmeg, dried rose petals, dry tea, cinnamon, herbs etc.

◊ A small piece of decorative fabric (of any type) for a cover, and matching thread

METHOD

◊ Cut two pieces of fabric in the shape you'd like. Add about 2 cm to the final dimensions all around, for the seam allowance.

◊ Pin the two pieces together, right sides facing.

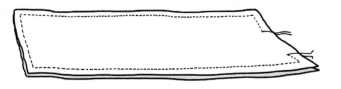

Figure 19.4: Sew around the edges of the fabric, leaving an opening at one end.

Figure 19.5: Trim the corners

◊ Sew the pieces together, leaving a gap of about 3 cm, where you can fill the bag (see Figure 19.4).

◊ Trim the corners as shown, and turn the bag inside out (see Figure 19.5).

◊ Using a funnel, fill the bag with your chosen material — uncooked rice is an easy and popular option.

◊ Pin the opening together, and sew it closed securely.

You might like to make a decorative cover for your heat pack, if the basic heat pack is in a plain cotton fabric. The cover can be in any fabric (as it won't be put in the microwave oven), and will help the pack to retain heat a bit longer. Simply make a cover a bit bigger than the dimensions of your heat pack. Sew three sides together, and hem the opening.

There are many tutorials online for different designs for heat packs you might like to try out, if you're feeling creative!

HOW TO USE

Heat it in your microwave oven in minute-long bursts and check, as there is the risk of it catching fire if microwaved for a long time! One or two minutes is usually plenty.

Easy peasy heat pack

If you don't want to be bothered with all that sewing, the easiest heat pack to make simply uses a long cotton sock and some uncooked rice. Put the rice into a long clean sock, until it's roughly ½ full, and tie a knot in the end. That's it!

Ice pack

Ice packs are the other tried and true item in any hippie's pain relief arsenal. While you can, of course, buy sturdy ice packs from your local pharmacy or supermarket, it is possible to make your own. In an emergency, never forget that old stand-by, a pack of frozen peas.

EQUIPMENT & MATERIALS

◊ 2 or 3 Ziplock plastic bags

◊ Water

◊ Cooking salt

METHOD

◊ Mix up some water with some salt. You can do this directly into the plastic bag, if you like, or use a jug. Roughly a 1:8 solution works well (i.e. ⅛ cup salt in 1 cup of water; or ¼ cup of salt in 2 cups of water).

◊ Two cups of salt solution is a good amount for a sandwich-sized Ziplock bag.

◊ Pour the salt mixture into the bag, and seal the bag, getting as much air out of the bag as you can in the process.

◊ Put the sealed bag upside down into the second bag (so the Ziplock seal is at the bottom of the second bag) and seal *that* bag. Go for a third bag for real security from leaks!

◊ Put the pack into the freezer until frozen.

Be aware that these packs may leak! It's best to use them for a short while, 20 minutes at most, and then put them *immediately* back into the freezer.

Don't let them sit for hours under the covers of your bed, slowly leaking into your mattress overnight (ask us how we know!). While you can add food colouring for fun, this also increases the clean up problems if it does leak. You have been warned!

Other Solutions that Work

◊ Undiluted dishwashing liquid

◊ Sugar solution (but this would be sticky if it leaks)

◊ Rubbing alcohol solution

We have seen some recipes for ice packs that use vodka, but we can think of much better things to do with that particular ingredient!

HOW TO USE

Adding substances like salt and alcohol to water lowers the water's freezing point As a result, the ice pack can get much colder than the usual 0° C of frozen water, yet still be flexible and mouldable to your body shape. It's important to avoid direct skin contact with any ice pack. Sports physician Dr Rob Reid recommends putting a dripping wet cloth between your skin and the pack when using it, to avoid skin damage. Limit the time the pack is on your skin to no more than 20 minutes at a time, three times a day.

Freja's Dressing Clips

This is a great dressing tool invented by HipWoman Freja Swogger. These clips are inexpensive and fast to make, and can be carried easily in your pocket or handbag — very useful for getting dressed at the swimming pool, when you can't bend down, for example!

Figure 19.6: Freja's Dressing Clips.

MATERIALS

◊ 4 large bulldog clips or 4 suspender clips

◊ Two shoelaces or long pieces of ribbon or thin cord, each piece should be around 1 metre long (4')

METHOD

◊ Tie one clip to each end of the two cords.

HOW TO USE

◊ Holding your pants (or trousers, or skirt), clip a clip to each side of the leg holes, on both sides.

◊ Hold onto the cords, and drop your pants to the floor.

◊ Step into the leg holes of your pants, and use the shoelaces to help pull them up!

◊ Remove the clips.

Walking stick and crutch decorations

If you're stuck with a walking stick and crutches, you may as well make them a feature! There are decorated walking sticks available now in some places, or you might like to go for a hand-carved special walking stick.

If, like most of us, you're stuck with plain old metal or wooden sticks, there are things you can do to make them truly unique!

Whatever you do to them, make sure you don't affect how they work, or their safety features. Don't paint over the clips that allow you to adjust the length of your crutches, for example.

STICKERS

The simplest and cheapest option is to use stickers. Choose some small stickers that you really like, and stick them along your crutches or walking stick, in a nice pattern. You can always remove them later if they get too grubby. Foil stickers will be more durable than paper stickers. Multiples of a single sticker design will look quite effective (a swarm of bees, perhaps?).

PAINT

You're only limited by your imagination here! The main thing is to prepare your surface properly, and use paints that are suitable for the material of your crutches (usually metal or wood).

Wooden crutches are easy to paint, but aluminium crutches aren't porous and the surface needs preparation so the paint stays on.

First of all, put down some protective newspaper or drop cloths on the floor. Wash, rinse, and dry your crutches. Cover or remove the areas you don't want to paint with masking tape or plastic bags (cover the padded armrests and remove the rubber foot tips, for example). Make sure any "adjustable" areas are protected; screws, sliding sections, holes/button latches, and so on. You don't want them gummed up with paint.

Wood
Sand lightly with fine sand paper to remove any outer treatment of the wood, and wipe them down with a damp cloth.

Metal
Spray the crutches with etching spray primer, and allow to dry (following the instructions on the can/tin). Don't use latex or acrylic primer on aluminium, or the final paint job will flake off.

Once the primer has dried, you can get started with the fun part!

Decorating Ideas
Use spray paint to make the crutches or walking stick a solid colour (how about fluorescent pink, or glow in the dark!?), or use a couple of colours to make soft stripes. Use a paint that is appropriate for use on metal; car spray paints are quite good for this.

Make sure you spray from a distance of at least 20 cm (8") to avoid drips and to get an even finish. Allow to dry thoroughly. Don't let the sticks dry lying down on the floor, as they might stick to the drop cloth — instead, prop them upright carefully. You can stop at this point.

If you like, add decorative elements to the sticks once the base colour has dried. Use little pots of enamel paint in your favourite colours, and a small brush. Dots, stripes, swirls, squares, flowers, ladybugs, or calligraphy are just a few ideas. You really are only limited by your imagination!

GOING TO TOWN!

There are tons of craft supplies out there that can be used on crutches and walking sticks. Decorative tapes are a fantastic option, as they stick on easily, create a great surface to put other things onto, and are easy to use, without any painty mess. Then you can use a hot glue gun to stick on whatever else you can imagine — beads, buttons, sequins — the options are practically endless.

The main thing to keep in mind is that the crutches are going to see some heavy duty work — getting around with you in rain, mud, and snow, for example — so you don't want anything on them that will be hard to clean (like feathers), or that impacts on the functioning of the crutches.

I must say, Bri's crutches were fabulous! I ordered holographic pink tape online and covered the crutches with it completely. For the underarm grips, we used zebra striped crutch cozies, which are also available online. And then, while she was in surgery for six hours, I superglued rhinestones all over them. She got so many positive comments on those crutches. She figured if she had to have them, she was going to do it in style. **Cynthia, 47 USA**

Knitted stick cosies

If you're a knitter, these little pull on or sew on crutch or walking stick cosies are a great way to add something a little special to your walking aids. They are very fast to make, and easy to remove if needed. You can, of course, adapt the patterns to make the same sort of thing with crochet.

MATERIALS & EQUIPMENT

◊ A small amount of sock yarn (4 ply)

◊ 2.5 mm (Size 1) needles, circular, double point needles (DPNs), or straights

◊ Darning needle

You can use different weight yarn with the appropriately sized needles (eg 8 ply/DK yarn and 4.0 mm (Size 6) needles). The main thing is you want to end up with a strip or tube that has a diameter of about 8 cm/3", which is what most walking sticks and crutches are. Any fancy stitch pattern would work here, too — lace, cables, and so on. Gauge isn't important.

WALKING STICKS

You can either knit a long thin rectangle and sew it into a tube, or knit it in a tube (knit in the round on 4 double-pointed needles, or on circular needles/DPNs).

Method

◊ Cast on about 24 stitches.

◊ Either knit back and forth in whatever stitch pattern you like or join in the round, and knit a tube.

◊ Knit to desired length.

◊ Bind off.

◊ Either sew onto the walking stick, or pull the tube up onto the stick.

CRUTCHES

Because of the hand rests and underarm sections, you won't be able to pull a knitted tube onto your crutches. You might like to just decorate the top section, or the whole crutch. These are knit flat, and then sewn onto the desired parts of the crutches.

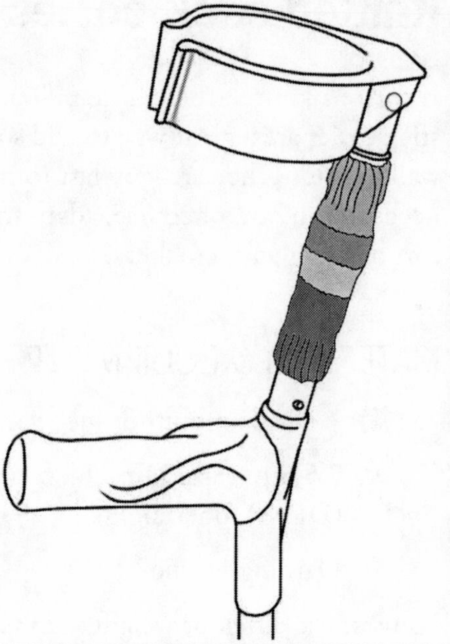

Method

◊ Using 4 ply yarn and 2.25 mm needles, cast on 24 sts.

◊ Work 8 rows of knit 1, purl 1 ribbing.

◊ Knit in stocking stitch to ~1 cm short of desired length.

◊ Work 8 rows of knit 1, purl 1 ribbing.

◊ Bind off.

◊ Sew onto crutch using mattress stitch.

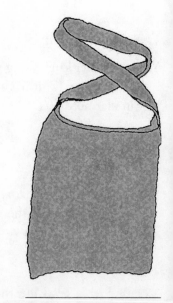

Figure 19.7: Elbow crutch with crutch cosy sewn on.

Knitted neck bag

This little bag is just the right size to sling about your neck and carry things in when you're on crutches, and can't use your hands to carry stuff.

The bag is knit flat and sewn up, but you could convert this to knitting in the round if you like. The Chinese Waves pattern is easiest

Figure 19.8. Knitted neck bag to use when on crutches.

knit flat, but you can adapt the pattern to have any stitch pattern you like, of course. Just avoid very lacy ones, as your things are more likely to fall through the holes.

Materials & Equipment

◊ 1 ball of 4 ply yarn, eg sock yarn or cotton

◊ Knitting needles 2.5 – 3.0 mm (Size 2 or 3) or so (a bit smaller than recommended for your yarn, to give a firmer fabric)

Figure 19.8 Chinese Waves

Chinese Waves Stitch Pattern

◊ Row 1: Repeat (Knit 1, slip 1) across the row

◊ Row 2: Knit

◊ Row 3: Knit 1, then repeat (Knit 1, slip 1) across the row (ie offset the slipped stitches)

◊ Row 4: Knit

In the slip stitch rows (#1 and #3), end each row with either k1 or k2, depending on how the slip stitches fall — you don't want the last stitch of the row to be a slipped stitch.

Bag

◊ Cast on 40 stitches.

◊ Knit in the Chinese Waves pattern as set until the piece is twice as long as you want the bag to be.

◊ Bind off.

◊ Fold in half, sew together side seams.

Neck strap

◊ Cast on 6 stitches.

◊ Knit it in garter stitch (knit every row).

◊ Bind off at desired length— keep in mind it will stretch quite a lot once used.

Finishing

Sew the ends of the neck strap firmly onto sides of bag near the opening. Darn in ends.

This pattern is accessible via Ravelry.com:

< www.ravelry.com/patterns/library/crutches-bag >

Photographing and printing x-rays

You may decide that after having surgery on your hips, you would like some evidence of this (other than the scars of course). Having an x-ray to show friends, family, boss, and colleagues can help them understand exactly what you have been through. Far easier than showing scars which, in the realms of DDH surgery, are in rather personal places!

There are several ways that you can get your x-ray, which depends on your hospital and your consultant. Ask your surgeon or their PA how to get hold of them, if you don't already have copies.

CD/DVD

If your x-rays are not automatically given to you (some hospitals will do this), many hospitals allow you to formally request a copy of your x-rays. You may have to fill in a form, and occasionally pay a fee, but usually there is no problem in getting them.

Once you have the disc, you can load the images onto your computer. Sometimes there is a special medical imaging program that needs to be loaded to view the images. If this is the case, it should be on the CD as well, and should automatically start when you click to load the images. Once the viewer is loaded, you can then save whichever images you want to your hard drive.

Memory/USB Stick

Sometimes you can give a memory stick to you surgeon, and they might be able to download the x-rays or scan images directly for you.

Film

Hard copies of x-rays and MRI scans on film are rarely seen in modern hospitals, as everything is done digitally nowadays. However, sometimes you may be given or see your images on films on light boards, and you may also have x-rays that you'd like to copy, from when you were younger.

If this is the case, it may be possible to simply take a photo of the pictures at hospital while they are on the light board (or even on your surgeon's computer screen).

At home, try holding or taping the x-rays to a sunny window or your brightly-lit computer screen, and photograph them that way. Basically, they need to be backlit. Scanning them on a scanner is difficult and not frightfully effective, even on "slide" setting.

To photograph them, use either a normal digital camera (set to maximum image quality), or a camera phone (which will generally be lower quality), while the x-ray is backlit. With this technique, the quality of the images will not be as good as having the digital copies, but you'll get images that are more than adequate for showing to others.

Once on your computer, you can upload them to Facebook, Myspace, HipWomen, your blog, or wherever you like. Or just keep them for your own satisfaction, and email them to friends and family.

Information cards

We have designed some little information cards which you can download from our web site < sutherland-studios.com.au >.

Simply print out the card that best applies to your situation, cut them out, fold them in half (they're double-sided), and keep a few in your wallet or purse. Hand them out when you don't feel like answering the endless questions!

There are cards for: developmental dysplasia of the hip, PAO, resurfacing, and THR.

Well, this is where we bid you *adieu*, and wish you all the very *very* best with your hip journey! We hope that this book helps you along the way, and that you're able to make friend with other hippies via groups like HipWomen, just as we've been able to do. It makes the whole experience a much less lonely one.

Sophie   Denise

Glossary

The first instance of each term within the book text is set in **bold**.

A few terms are not found in this book, but are used by medical professionals when discussing hip dysplasia.

Abbreviations are listed on page viii at the front of the book.

abduct to move away from the midline of the body

abductor a muscle that, when contracted, moves a limb away from the body

acetabular labrum rim of cartilage surrounding the hip socket that acts to deepen it

acetabulum the hip socket

adduct to move towards the midline of the body

adductor a muscle that, when contracted, moves a limb towards the body

aetiology the set of causes of a disease or condition

allograft tissue or bone donation from another person

anaemia too little haemoglobin in the blood

analgesia the medical term for pain -relieving medications

anaphylaxis rapid and severe, life-threatening, allergic reaction

anticoagulant a drug or substance that thins the blood

antipyretic a drug or substance that decreases a fever or the body temperature

arthroplasty replacement of a joint with a man-made one

arthrogram a radiographic scan of a joint, after the injection of a contrast dye into the joint

arthroplasty surgical reconstruction or replacement of a malformed or arthritic joint

arthroscopy keyhole surgery on a joint, where instead of a big incision, several much smaller ones are made. Then, using special cameras and instruments, the surgeon does the procedure using a video screen to see inside the joint

articular cartilage the soft, smooth substance that covers the parts of bones that form a joint and provides a cushioning effect

aseptic free from contamination by bacteria, viruses, or other infectious agents

autologous obtained from one's self, to be used by one's self, eg autologous blood donation

avascular necrosis bone death caused by poor blood supply, which can affect the head of the femur

Barlow test a test done on babies in the first few months of life to test if the hip is stable or unstable

bilateral occurring on both sides of the body

biocompatible compatible with living cells, and not causing any injury, toxicity, or rejection by the body

breech birth when a baby is born buttocks or feet first

bursa fluid-filled sac that protects areas of the body against friction (such as where a muscle rubs against a bone)

bursitis inflammation of a bursa

cartilage tough, smooth, elastic connective tissue found in the joints, outer ear, nose, and larynx

congenital acetabular dysplasia (CAD) a type of hip dysplasia where the femoral head and neck are normal

cell saver a machine that takes the blood lost during surgery then cleans and filters it so it can be given directly back to you

claustrophobic fearful of being in narrow or enclosed spaces

closed reduction a procedure done in babies and infants where the hip joint is relocated and then held in place with a special plaster cast

coccyx small triangular bone at the base of the spine, the 'tailbone'

congenital a disease or abnormality present from birth

computed tomography (CT) scan imaging method using a special x-ray scanner that creates cross-section images of the body

congenital dysplasia/dislocation of the hip (CDH) a condition where the hip joint does not develop normally and comprises a spectrum from the completely dislocated hip to those which are unstable but in the correct place

cognitive behaviour therapy (CBT) a type of psychological therapy that teaches you to perceive and see things differently. It can be used to help with depression, anxiety, and pain

coxa valga a deformity of the femur where the femoral neck is at a bigger angle than normal (+125°) in relation to the shaft of the femur

coxa vara a deformity of the femur where the femoral neck is at a smaller angle than normal (-125°) in relation to the shaft of the femur

coxodynia a painful coccyx or tail bone

cytokines proteins, which are not antibodies, released by cells as part of the immune response

debridement removal of dead or damaged tissue and debris from a wound or damaged area of the body

deep vein thrombosis (DVT) a clot of blood that forms in the deep veins of the leg, treated with medication that thins the blood

developmental dysplasia of the hip (DDH) a condition where the hip joint does not develop normally and comprises a spectrum from the completely dislocated hip to those which are unstable but in the correct place

directed/designated donation the donation of blood for a designated recipient

dosset box a box used to organise medication, with individual slots for each day

dysplasia abnormality of development

electrocardiogram (ECG/EKG) a tracing of the electrical activity of the heart

epidemiology the study of the incidence, distribution, and control of diseases and conditions in populations

excision surgical removal of something by cutting

external rotation of the hip when the hip is straight and in line with the body, the knee and toes are rolled to be pointing away from the midline (duck-footed)

extra-articular outside the joint capsule

femoral head the ball part at the top of the femur that forms a joint with the acetabulum (hip socket)

femoral acetabular impingement (FAI) a condition where the femoral head rubs against the socket abnormally

femur the thigh bone

flexion the action of bending a joint, or the condition of a joint being in a bent position

flexor a muscle that flexes or bends a joint

fluoroscopy an x-ray procedure that makes it possible to see real-time video images

general anaesthesia (GA) complete unconsciousness and insensitivity to pain caused by the administration of special drugs, used during most surgeries

genetic a disorder or condition that is caused by an abnormal gene; this can be inherited or a spontaneous mutation (i.e. new in that person)

gluteus any of the three muscles forming the buttocks, that extend and turn the thigh

greater trochanter the large bony prominence just below and on the outside of the femoral neck. It is the bony part that can be felt at the top of your thigh

haemoglobin the oxygen-carrying molecule in blood

hereditary determined by genetic factors, and able to be passed down from parents to child

hip dysplasia common name for developmental dysplasia of the hip (DDH)

hip restrictions/precautions a set of limitations to normal movement of the hip, which much be adhered to after major hip surgery (especially hip replacement), while the joint capsule and muscles heal. The precautions guard against dislocation and joint damage

hip spica a special type of plaster cast that holds the hips and legs in a certain position

hypermobile abnormally wide range of motion in joints

hypertrophic increase in size of an organ or tissue because of the enlargement of its cells

iliacus a muscle that arises from the inner surface of the pelvis and acts to flex the hip joint

iliopsoas group of muscles of the inner hip, or this general region

ilium a bone that forms part of the pelvis

impingement where tissue is compressed or pinched by movements of a joint, or bone hits on bone in an abnormal way

in utero inside the uterus; occuring before birth

incidence a measure of the risk of developing a condition within a certain time period

intensive care unit (ICU), intensive treatment/therapy unit (ITU) a special hospital ward where patients who are very sick or who require special monitoring are looked after. There is a high ratio of nurses to patients, usually 1:1

internal rotation of the hip when the leg is straight and in line with the body, the hip is rolled so the knee and foot point towards the midline (pigeon-toed)

international normalised ratio (INR) a measure of how thin the blood is. Normal is 1.0, abnormal is any number greater than 1.0

intra-articular inside the joint capsule

intravenous (IV) where a substance is given directly into a vein. This can be drugs, fluid, or blood

iontophoresis use of an electrical current to move ions of a medicine (such as cortisone) through the skin, into the body

ischium a bone of the pelvis

isometric exercise strength training exercise performed against unmoving resistance, typically muscle contractions, with static body positions (eg weight lifting)

joint (articular) capsule saclike envelope enclosing a joint, made of fibrous tissues

keyhole (arthroscopic/ laparoscopic) surgery minimally invasive surgical technique, where small incisions are made, and special cameras and instruments are used

labrum a rim of cartilage that surrounds the hip or shoulder socket and helps to stabilise it

lateral centre edge angle this is a measurement of how shallow a hip socket is in relation to the femoral head, normal is 25 degrees; less than 20 degrees indicates dysplasia

laxative a drug or substance that induces or increases the frequency of bowel motions or softens the stool

laxity slackness or looseness

lesser trochanter the smaller bony prominence just below the neck of the femur

ligament a fibrous structure that connects bone to bone

ligamentum teres a strong round ligament that attaches the head of the femur to the inside of the acetabulum

local anaesthesia the numbing of a region through the injection of special drugs; the patient remains conscious during this type of anaesthesia

loose body fragment of cartilage or other solid tissue floating within a joint

macrophage large white blood cell in the blood and connective tissues that ingests disease-causing organisms

magnetic resonance imaging (MRI) scan a type of scan that uses alternating magnetic fields to visualise the structures of the body

muscle a bunch of fibrous tissues in the body that contract and relax to move parts of the body. They are connected to the bones by tendons

neuropathic pain originating from damage to nerves

nil by mouth (NBM) not being allowed to have anything to eat or drink

nonunion failure of bones to heal together after fracture or osteotomy

non-weight-bearing (NWB) no body weight is to be supported by the operated leg

oedema an abnormal excess accumulation of fluid within the body

oligohydramnios to little fluid in the amniotic sac during pregnancy

open reduction an operation done on young children that requires an incision and then the hip joint bones are manually put into their normal position (reduced)

opiod/opiate a pain-killing drug derived from opium

orthopaedic surgeon (OS) a surgeon who specialises in diagnosing and treating problems of the musculoskeletal system

orthopaedic surgery to do with correcting problems of the musculoskeletal system

orthosis a brace, splint, or other artificial external device which supports the limbs or spine, or prevents or helps relative movement

Ortolani test a test done on babies in the first few months of life to test if the hip is stable or unstable

osteoarthritis (OA) a type of arthritis where the cartilage lining of a joint erodes away leading to pain, stiffness, and loss of movement

osteolysis the destruction or disappearance of bone tissue caused by a disease process

osteophyte a small abnormal bony growth or spur, associated with osteoarthritis

paediatric the branch of medicine dealing with children

pathology the study of disease processes

patient-controlled analgesia (PCA) intravenous pain relief controlled by the patient

perineum diamond-shaped region of the human trunk between the legs, between the anus and urogenital passages

pelvis large ring-shaped bony structure that rests on the legs, and supports the spine

periacetabular osteotomy (PAO) an operation where the bones of the pelvis are broken around the acetabulum which is then moved and repositioned to create a better and more stable hip joint

perioperative time period of a surgical procedure, including admission, surgery, and recovery

physiotherapy/physical therapy (PT) the healthcare profession associated with maximising quality of life through rehabilitiation, exercises, massage, electrotherpy, and patient education and training

platelets the part of blood involved in clotting and immune functions

post-anaesthesia care unit (PACU) a special ward specifically for patients post-surgery who require more intensive monitoring or nursing

post-op (post-operative) after surgery

post-traumatic stress disorder (PTSD) a psychological condition precipitated by previous trauma in one's life, this may by physical, emotional, or psychological trauma

pre-op (pre-operative) before surgery

prevalence the proportion of people with a disease out of all the population

prosthesis an artificial body part

psoas major a muscle that runs from the lower spine to the femur, and helps to move the hip joint

pubis bone that forms part of the pelvis

pulmonary embolism (PE) a blood clot in the lung's blood vessels

quadriceps large muscle group on the front of the thigh

range of motion (ROM) distance and direction a joint can travel

rectus femoris a muscle that runs from the pelvis to just below the knee, responsible for knee extension and hip flexion

regional anaesthesia a type of anaesthesia where a region of the body is numbed

resurfacing a surgery where just the surfaces of the hip joint are replaced

revision THR the replacement of an old worn-out hip prosthesis with a new one

sacrum large, concave, triangular bone at the end of the spine

sciatic nerve longest and widest nerve in the human body, running from the lower back to the lower limbs

sign an objective finding (e.g. x-ray or test) showing the existence of disease, observable by a doctor, but not always obvious to the patient

sock gutter a disability aid that helps to put socks on

soft tissues tissues in the body that are not bone e.g. muscles, tendons, nerves, and ligaments

symptom subjective indication of a disease, discernable by the patient (eg pain, fever, nausea)

synovial a type of joint which is surrounded by a thick flexible membrane forming a sac into which is secreted a thick fluid that lubricates the joint

tendon a strong flexible cord of fibrous tissue that attaches muscles to bones

tendonectomy a surgical procedure where a tendon is cut

tendonitis swelling, inflammation, and irritation of a tendon

tibia one of the leg bones between the knee and ankle

toe-touch weight-bearing the foot or toes may touch the floor, just to maintain balance, but not support any body weight

total hip replacement (THR) surgical procedure where a worn out or damaged hip joint is removed, and an artificial hip joint is put in place

traction a force placed on a body part to maintain a bone or joint in a certain position

transdermal through or via the skin

trochanter either of two bony protuberances on the top of the femur, where the thigh muscles are attached

unilateral occurring on only one side of the body

weight-bearing zone region of a joint that supports the weight of the body. In a normal hip this zone is flat, in a dysplastic hip it is angled

World Health Organisation (WHO) a specialised body of the United Nations that is a coordinating body on international health

Recommended Reading
and Listening

◊ *Full Catastrophe Living* by Jon Kabat-Zinn

◊ *Heal Your Hips : How to Avoid Hip Surgery — and What to Do If You Need It* by Robert Klapper and Lynda Huey

◊ *If Not Dieting, Then What?* by Rick Kausman

◊ *Manage Your Pain* by Michael Nicholas, Allan Molloy, Lois Tonkin, and Lee Beeston

◊ *Mindfulness for Beginners* (audio) by Jon Kabat-Zinn

◊ *Mindfulness Skills* (audio) by Russ Harris

◊ *Reinventing Your Life : The Breakthrough Program to End Negative Behavior ... and Feel Great Again* by Jeffrey E. Young and Janet S. Klosko

◊ *Sick and Tired of Being Sick and Tired: Living with Invisible Chronic Illness* by Paul J. Donoghue and Elizabeth Siegel

◊ *The Anxiety and Phobia Workbook* by Edmund J Bourne

◊ *The Demon-Haunted World* by Carl Sagan

◊ *The Parents' Guide to Hip Dysplasia* by Betsy Miller

◊ *The Relaxation and Stress Reduction Workbook* by Martha Davis, Elizabeth Eshelman, and Matthew McKay

◊ *Trick or Treatment* by Edzard Ernst and Simon Singh

Cited References

1. Troelsen A, 'Surgical advances in periacetabular osteotomy for treatment of hip dysplasia in adults', *Acta Orthopaedica Supplement*, April 2009, 80(332):1-33.

2. Parvizi J, Bican O, Bender B, Mortazavi SM, Purtill JJ, Erickson J, et al. 'Arthroscopy for labral tears in patients with developmental dysplasia of the hip: a cautionary note', *The Journal of Arthroplasty*, Sept 2009, 24(6 Suppl):110-3.

3. Ali AM, Angliss R, Fujii G, Smith DM, Benson MK, 'Reliability of the Severin classification in the assessment of developmental dysplasia of the hip', *Journal of Pediatric Orthopaedics*, Oct 2001,10(4):293-7.

4. Sewell MD, Rosendahl K, Eastwood DM, 'Developmental dysplasia of the hip', *British Medical Journal*, 2009, 339:b4454.

5. Lewis K, Jones DA, Powell N, 'Ultrasound and neonatal hip screening: the five-year results of a prospective study in high-risk babies', *Journal of Pediatric Orthopaedics*, Nov-Dec 1999,19(6):760-2.

6. Stevenson DA, Mineau G, Kerber RA, Viskochil DH, Schaefer C, Roach JW, 'Familial predisposition to developmental dysplasia of the hip', *Journal of Pediatric Orthopaedics*, Jul-Aug 2009, 29(5):463-6.

7. Witt JW, 'Hip Dysplasia', < www.hipjointsurgery.co.uk >, accessed November 2011.

8. Jacobsen S, 'Adult hip dysplasia and osteoarthritis. Studies in radiology and clinical epidemiology', *Acta Orthopaedica Supplement*, Dec 2006, 77(324):1-37.

9. 'Radiation Exposure in X-ray and CT Examinations', *Radiological Society of North America*, < www.radiologyinfo.org/en/safety/index.cfm?pg=sfty_xray >, accessed November 2011.

10. Garbuz DS, Masri BA, Haddad F, Duncan CP, 'Clinical and radiographic assessment of the young adult with symptomatic hip dysplasia', *Clinical Orthopaedics and Related Research*, 2004 Jan(418):18-22.

11. Seror P, Pluvinage P, d'Andre FL, Benamou P, Attuil G. 'Frequency of sepsis after local corticosteroid injection (an inquiry on 1,160,000 injections in rheumatological private practice in France)', *Rheumatology* (Oxford), 1999 Dec;38(12):1272-4.

12. Kumar N, Newman RJ, 'Complications of intra- and peri-articular steroid injections', *The British Journal of General Practice*, 1999 Jun;49(443):465-6.

13. Wang CT, Lin J, Chang CJ, Lin YT, Hou SM, 'Therapeutic effects of hyaluronic acid on osteoarthritis of the knee. A meta-analysis of randomized controlled trials', *The Journal of Bone and Joint Surgery, American Volume*, 2004 Mar;86-A(3):538-45.

14. Wheeless, CR, 'Hip Joint Arthroscopy', *Wheeless' Textbook of Orthopaedics* (online), Duke Orthopaedics; <a www.wheelessonline.com/ortho/hip_joint_arthroscopy >, accessed November 2011.

15. Shetty VD, Villar RN, 'Hip arthroscopy: current concepts and review of literature', *British Journal of Sports Medicine*, [Review], 2007;41:64-8.

16. Dorrell JH, Catterall A, 'The torn acetabular labrum', *The Journal of Bone and Joint Surgery. British Volume*, 1986 May;68(3):400-3.

17. Klaue K, Durnin CW, Ganz R, 'The acetabular rim syndrome. A clinical presentation of dysplasia of the hip', *The Journal of Bone and Joint Surgery. British Volume,*1991 May;73(3):423-9.

18. Estes C, Taunton M, 'Hip Dysplasia', *Orthopaedia*, < www.orthopaedia.com/display/ Main/Hip+dysplasia > accessed November 2011.

19. Clohisy JC, Schutz AL, St John L, Schoenecker PL, Wright RW, 'Periacetabular osteotomy: a systematic literature review', *Clinical Orthopaedics and Related Research*, 2009 Aug;467(8):2041-52.

20. van Bergayk AB GD, 'Quality of life and sports-specific outcomes after Bernese periacetabular osteotomy', *The Journal of Bone and Joint Surgery. British Volume*, 2002;84-B(3):339-43.

21. Biedermann R, Donnan L, Gabriel A, Wachter R, Krismer M, Behensky H, 'Complications and patient satisfaction after periacetabular pelvic osteotomy', *International Orthopaedics*, 2008 Oct;32(5):611-7.

22 Matheney T, Kim YJ, Zurakowski D, Matero C, Millis M, 'Intermediate to long-term results following the Bernese periacetabular osteotomy and predictors of clinical outcome', *The Journal of Bone and Joint Surgery, American Volume*, 2009 Sep;91(9):2113-23.

23. Steppacher SD, Tannast M, Ganz R, Siebenrock KA, 'Mean 20-year followup of Bernese periacetabular osteotomy', *Clinical Orthopaedics and Related Research*, 2008 Jul;466(7):1633-44.

24. Sambandam SN, Hull J, Jiranek WA, 'Factors predicting the failure of Bernese periacetabular osteotomy: a meta-regression analysis', *International Orthopaedics*, 2009 Dec;33(6):1483-8.

25. Millis MB, Kain M, Sierra R, Trousdale R, Taunton MJ, Kim YJ, et al. 'Periacetabular osteotomy for acetabular dysplasia in patients older than 40 years: a preliminary study', *Clinical Orthopaedics and Related Research*, 2009 Sep;467(9):2228-34.

26. Matta JM, Stover MD, Siebenrock K, 'Periacetabular osteotomy through the Smith-Petersen approach', *Clinical Orthopaedics and Related Research*, 1999 Jun(363):21-32.

27. Trumble SJ, Mayo KA, Mast JW, 'The periacetabular osteotomy. Minimum 2 year follow up in more than 100 hips', *Clinical Orthopaedics and Related Research*, 1999 Jun(363):54-63.

28. Clohisy JC, Nunley RM, Curry MC, Schoenecker PL, 'Periacetabular osteotomy for the treatment of acetabular dysplasia associated with major aspherical femoral head deformities', *The Journal of Bone and Joint Surgery. American Volume*, 2007 Jul;89(7):1417-23.

405

29. Peters CL, Erickson JA, Hines JL, 'Early results of the Bernese periacetabular osteotomy: the learning curve at an academic medical center', *The Journal of Bone and Joint Surgery. American Volume,* 2006 Sep;88(9):1920-6.

30. Pulido LF, Babis GC, Trousdale RT, 'Rate and risk factors for blood transfusion in patients undergoing periacetabular osteotomy', *Journal of Surgical Orthopaedic Advances,* 2008 Fall;17(3):185-7.

31. Yeung. E RA, Witt. J, 'Blood transfusion requirements in patients undergoing peri-acetabular osteotomy for hip dysplasia', *Journal of Bone and Joint Surgery, British Volume,* Vol 88-B, Issue SUPP_II, 245-246, 2005;88-B(Supplement II):245-6.

32. Trousdale RT, Ekkernkamp A, Ganz R, Wallrichs SL, 'Periacetabular and intertrochanteric osteotomy for the treatment of osteoarthrosis in dysplastic hips', *The Journal of Bone and Joint Surgery. American Volume,*1995 Jan;77(1):73-85.

33. Xu WD, Li J, Zhou ZH, Wu YS, Li M, 'Results of hip resurfacing for developmental dysplasia of the hip of Crowe type I and II', *Chinese Medical Journal (English edition),* 2008 Aug 5;121(15):1379-83.

34. Li J, Xu W, Xu L, Liang Z, 'Hip resurfacing for the treatment of developmental dysplasia of the hip', *Orthopedics,* 2008 Dec;31(12).

35. Naal FD, Schmied M, Munzinger U, Leunig M, Hersche O, 'Outcome of hip resurfacing arthroplasty in patients with developmental hip dysplasia', *Clinical Orthopaedics and Related Research,* 2009 Jun;467(6):1516-21.

36. Baque F, Brown A, Matta J, 'Total hip arthroplasty after periacetabular osteotomy', *Orthopedics,* 2009 Jun;32(6):399.

37. Biant LC, Bruce WJ, Assini JB, Walker PM, Walsh WR, 'Primary total hip arthroplasty in severe developmental dysplasia of the hip. Ten-year results using a cementless modular stem', *The Journal of Arthroplasty,* 2009 Jan;24(1):27-32.

38. Santori FS, Vitullo A, Stopponi M, Santori N, Ghera S, 'Prophylaxis against deep-vein thrombosis in total hip replacement. Comparison of heparin and foot impulse pump', *The Journal of Bone and Joint Surgery. British Volume,*1994 Jul;76(4):579-83.

39. Comp PC, Spiro TE, Friedman RJ, Whitsett TL, Johnson GJ, Gardiner GA, Jr., et al. 'Prolonged enoxaparin therapy to prevent venous thromboembolism after primary hip or knee replacement. Enoxaparin Clinical Trial Group', *The Journal of Bone and Joint Surgery. American Volume,* 2001 Mar;83-A(3):336-45.

40. Shen B, Yang J, Wang L, Zhou ZK, Kang PD, Pei FX, 'Midterm results of hybrid total hip arthroplasty for treatment of osteoarthritis secondary to developmental dysplasia of the hip-Chinese experience', *The Journal of Arthroplasty,* 2009 Dec;24(8):1157-63.

41. Togrul E, Ozkan C, Kalaci A, Gulsen M, 'New Technique of Subtrochanteric Shortening in Total Hip Arthroplasty for Crowe Types 3 to 4 Dysplasia of the Hip Using Endosteal Bone Pegs', *The Journal of Arthroplasty,* 2009 July 3.

42. 'Aseptic Loosening of Total Hip Replacement', *Orthopaedia,* < www.orthopaedia.com/display/Review/Aseptic+Loosening+of+Total+Hip+Replacement >, accessed August 2010.

43. Surin, HBV, 'More Details on Aseptic Loosening of Total Hips', *Facts about Total Joints*, < www.newtotaljoints.info/TH_loose_DETAILS.htm >, accessed April 2010.

44. Jager M, Endres S, Wilke A, 'Total hip replacement in childhood, adolescence and young patients: a review of the literature', *Zeitschrift für Orthopädie und ihre Grenzgebiete*, 2004 Mar-Apr;142(2):194-212.

45. Jager M, Begg MJ, Ready J, Bittersohl B, Millis M, Krauspe R, et al. 'Primary total hip replacement in childhood, adolescence and young patients: quality and outcome of clinical studies', *Technology and Health Care*, 2008;16(3):195-214.

46. Porsch M, Siegel A, 'Artificial hip replacement in young patients with hip dysplasia--long-term outcome after 10 years', *Zeitschrift für Orthopädie und ihre Grenzgebiete*, 1998 Nov-Dec;136(6):548-53.

47. Engesaeter LB, Furnes O, Havelin LI, 'Developmental dysplasia of the hip--good results of later total hip arthroplasty: 7135 primary total hip arthroplasties after developmental dysplasia of the hip compared with 59774 total hip arthroplasties in idiopathic coxarthrosis followed for 0 to 15 years in the Norwegian Arthroplasty Register', *The Journal of Arthroplasty*, 2008 Feb;23(2):235-40.

48. Mahomed NN, Barrett JA, Katz JN, Phillips CB, Losina E, Lew RA, et al. 'Rates and outcomes of primary and revision total hip replacement in the United States medicare population', *The Journal of Bone and Joint Surgery. American Volume*, 2003 Jan;85-A(1):27-32.

49. Barbosa-Cesnik C, Brown MB, Buxton M, Zhang L, DeBusscher J, Foxman B, 'Cranberry juice fails to prevent recurrent urinary tract infections', *Clinical Infectious Diseases*, 2011 Jan 1;52(1): 23-30.

50. Jacobsen S and Sonne-Holm S, 'Hip dysplasia: a significant risk factor for the development of hip osteoarthritis. A cross-sectional survey', *Rheumatology*, 2005 44(2):211-218; doi:10.1093/rheumtology/keh436.

51. Hollis JF, et al. 'Weight Loss During the Intensive Intervention Phase of the Weight-Loss Maintenance Trial', *American Journal of Preventative Medicine*, Volume 35, Issue 2, Pages 118-126, < www.ajpmonline.org/article/S0749-3797(08)00374-7/abstract > August 2008.

52. Babyak M, et al. 'Exercise Treatment for Major Depression: Maintenance of Therapeutic Benefit at 10 Months', *Psychosomatic Medicine* 62:633–638 (2000).

53. Baer RA, 'Mindfulness Training as a Clinical Intervention: A Conceptual and Empirical Review', *Clinical Psychology: Science and Practice*, 10(2), 125-143.

54. Wiseman, R, *59 Seconds. Think a little, change a lot*. Alfred A. Knopf, 2009, pg 196.

55. Lilienfeld SO, SJ, Ruscio J, Beyerstein BL, *50 Myths of Popular Psychology*, Wiley-Blackwell, 2010, pg 129-132.

56. Smyth JM, Stone AA, Hurewitz A, Kaell A, 'Effects of Writing About Stressful Experiences on Symptom Reduction in Patients With Asthma or Rheumatoid Arthritis : A Randomized Trial', *The Journal of the American Medical Association*, 1999;281(14):1304-1309. doi: 10.1001/jama.281.14.1304.

57. Baikie KA, Kay Wilhelm K, 'Emotional and physical health benefits of expressive writing', *Advances in Psychiatric Treatment*, (2005) 11: 338-346

58. Young KD, 'Pediatric procedural pain', *Annals of Emergency Medicine*, 2005; 45:160-171.

59. Evans RG, *Paediatrics in New South Wales, 1945 to 1965*, PhD thesis, University of Newcastle, December 2000.

60. Markel H, 'When Hospitals Kept Children from Parents', *New York Times*, Jan 08 < www.nytimes.com/2008/01/01/health/01visi.html >, accessed August 2010.

61. 'Healing Emotional and Psychological Trauma' < helpguide.org/mental/emotional_psychological_trauma.htm >, accessed November 2011.

62. 'Pain, Pain, Go Away', *American Psychological Association* < www.apa.org/research/action/pain.aspx >, accessed February 2011.

63. Donoghue P, Siegel M, *Sick and Tired of Feeling Sick and Tired : Living with Invisible Chronic Illness*, W. W. Norton Company, New York, 1992. pp 89-94.

64. Mundy GM, Birtwistle SJ, Power RA, 'The effect of iron supplementation on the level of haemoglobin after lower limb arthroplasty', *The Journal of Bone and Joint Surgery. British Volume*, 2005 Feb;87(2):213-7.

65. Moller AM, Pedersen T, Villebro N, Munksgaard A, 'Effect of smoking on early complications after elective orthopaedic surgery', *The Journal of Bone and Joint Surgery. British Volume*, 2003 Mar;85(2):178-81.

66. Moores LK, 'Smoking and postoperative pulmonary complications. An evidence-based review of the recent literature', *Clinics in Chest Medicine*, 2000 Mar;21(1):139-46, ix-x.

67. Moller AM, Pedersen T. 'The effect of tobacco smoking on risks in connection in anesthesia and surgery. Development of complications and the preventive effect of smoking cessation', *Ugeskrift for Laeger*, 1999 Jul 26;161(30):4273-6.

68. Sadr Azodi O, Lindstrom D, Adami J, Tonnesen H, Nasell H, Gilljam H, et al. 'The efficacy of a smoking cessation programme in patients undergoing elective surgery: a randomised clinical trial', *Anaesthesia*, 2009 Mar;64(3):259-65.

69. 'Will I need a blood transfusion?' *National-Blood-Service*, <www.blood.co.uk/pdf/Final_Version_Will_I_Need.pdf >, accessed October 2011.

70. Mundy GM, Birtwistle SJ, Power RA, 'The effect of iron supplementation on the level of haemoglobin after lower limb arthroplasty', *The Journal of Bone and Joint Surgery. British Volume*, 2005 Feb;87(2):213-7.

71. Barclay L, 'Daily Calcium Plus Vitamin D Supplements May Reduce Fracture Risk', *British Medical Journal*, 2010;340:b5463.

72. Cheng B, Hung CT, Chiu W, 'Herbal Medicine and Anaesthesia', *Hong Kong Medical Journal*, Vol 8 No 2 April 2002 pp 123-130.

73. Rowe DJ, Baker AC, 'Perioperative risks and benefits of herbal supplements in aesthetic surgery', *Aesthetic Surgery Journal*, 2009 Mar-Apr;29(2):150-7.

74. 'Clinical Guidelines for Contraceptive Use', *Faculty of Sexual and Reproductive Healthcare, Royal College of Obstetricians and Gynaecologists*, < www.fsrh.org/pages/clinical_guidance.asp > Various documents, accessed November 2011.

75. Davey JP, Santore RF, 'Complications of periacetabular osteotomy', *Clinical Orthopaedics and Related Research*, 1999 Jun(363):33-7.

76. Walter WL, Yeung E, Esposito C, 'A Review of Squeaking Hips', *The Journal of the American Academy of Orthopaedic Surgeons*, Vol 18, No 6, June 2010, 319-326.

77. Downes EM, 'Late Infection after Total Hip Replacement', *The Journal of Bone and Joint Surgery*, Feb 1977, Vol 59-B, 1 : 42-44.

78. Surin HBV, 'Treatment of the Total Hip Infection', *Facts about Total Joints*, < www.newtotaljoints.info/TREAT_HIPINFECT.htm > accessed July 2011.

79. 'Total Hip Replacement', *Your Orthopaedic Connection, American Academy of Orthopedic Surgeons*, < orthoinfo.aaos.org/topic.cfm?topic=A00377 > accessed September 2010.

80. Sinha RK, Shanbhag AS, Maloney WJ, Hasselman CT, Rubash HE, 'Osteolysis: Cause and Effect', *Instructional Course Lectures*, Volume 47. Rosemont, Ill: American Academy of Orthopaedic Surgeons Press; 1998: 307-320.

81. Furnes O, Lie SA, Espehaug B, Vollset SE, Engesaeter LB, Havelin LI, 'Hip disease and the prognosis of total hip replacements. A review of 53,698 primary total hip replacements reported to the Norwegian Arthroplasty Register 1987-99', *The Journal of Bone and Joint Surgery. British Volume*, 2001 May;83(4):579-86.

82. Boynton E, Waddell JP, Morton J, Gardiner GW, 'Aseptic loosening in total hip implants: the role of polyethylene wear debris', *Canadian Journal of Surgery. Journal Canadien de Chirurgie*, 1991 Dec;34(6):599-605.

83. 'Aseptic Loosening', *Orthopaedic Web Links*, < www.orthopaedicweblinks.com/Detailed/9573.html >, accessed January 2011.

84. Oransky I, 'The Effects of Osteolysis and Aseptic Loosening', *Medscape Orthopaedics Sports Medicine*, 2001;5(1).

85. Bordini B, Stea S, De Clerico M, Strazzari S, Sasdelli A, Toni A, 'Factors affecting aseptic loosening of 4750 total hip arthroplasties: multivariate survival analysis', *BioMed Central Musculoskeletal Disorders* (2007) Volume: 8, Issue: 1471-2474 (Electronic)

86. Wheeless CR, 'Dislocation of THA', *Wheeless' Textbook of Orthopedics*, < www.wheelessonline.com/ortho/dislocation_of_tha >, accessed February 2011.

87. Surin HBV, 'Disloaction of the Total Hip', *Facts about Total Joints*, < www.newtotaljoints.info/DISLOCARION_totalhip.htm>, accessed July 2011.

88. Valenzuela RG, Cabanela ME, Trousdale RT, 'Sexual Activity, Pregnancy, and Childbirth After Periacetabular Osteotomy', *Clinical Orthopaedics Related Research*, January 2004 - Volume 418 - Issue - pp 146-152.

89. Masui T, et al. 'Childbirth and sexual activity after eccentric rotational acetabular osteotomy,' *Clinical Orthopaedics Related Research*, 2007 Jun;459:195-9.

Index

419

workplaces (continued)
    sick pay, 67
    time off for surgery, 214
    work, in hospital, 227
worries about surgery, 169–172
wound care, 287–288, 307
wound drains, 239
wound infections, 74–75, 288
    related to smoking, 190

X

x-rays (illustrations)
    dysplasic hip, 3
    normal hip, 3
    periacetabular osteotomy, 48
    resurfacing, 56
    total hip replacement, 56
x-rays (imaging technique), 21–22
    to detect aspectic loosening, 316
    for diagnosing osteoarthritis, 79
    difficulties interpreting, 2
    and embarrassment, 21
    as guidance for joint injections, 23, 39
    how to copy, 392–393
    of infant hips, 6
    before pregnancy, 325
    radiation levels, 21
    sharing online, 346

Y

Yahoo Groups (website), 344–346
yoga, 31
Young, Jeffrey, 117
young adults with dysplasia, 139–144

CPSIA information can be obtained at www.ICGtesting.com
Printed in the USA
LVOW080952170712

290400LV00001B/133/P